AF577412

Pentagon's

MILITARY QUOTES

Pentagon's MILITARY QUOTES

Compiled by

Col Harjeet Singh (Retd)

Pentagon's Military Quotes / Col Harjeet Singh (Retd)

ISBN 978-81-8274-565-0

First Published in 2012

Copyright © Reserved

Published by

PENTAGON PRESS
206, Peacock Lane
Shahpur Jat, New Delhi-110049
Phones: 011-64706243, 26491568
Telefax: 011-26490600
email: rajan@pentagonpress.in
website: www.pentagonpress.in

All rights reserved. No part of this publication may be copied, reproduced, stored in a retrieval system or transmitted in any form or by any means, electronic, mechanical, photocopying, recording, or otherwise, without the prior permission of the publisher.

Printed at Aegean Offset Printers, Greater Noida, U.P.

INTRODUCTION

Researching for this book on military quotes has been an enriching and erudite experience. The subject is vast. History has a long and venerable pedigree and in the recorded history of man there has never been a century without a war. Mankind has had ten-thousand years of experience at fighting in various forms of combat, and the literature on the subject is enormous. The choice of quotes has perforce to be eclectic. The publisher, Rajan Arya, while being very supportive, had laid down a page limit. This constrained its own brevity, especially when one felt that there was always one more quote to include. Be that as it may, the endeavour has been to include quotes on every facet of the military and those that impact it in various ways.

Quotations inspire us, make us laugh, give us perspective, and sometimes just help us say what needs to be said.

There are different categories of quotes in this volume – the military profession encompasses duty, honour, courage, valour, humour, ethos, experience of soldiering, knowledge of history, inspiration, motivation, management, command and leadership, amongst myriad issues of strategy and tactics which are representative of the military and the fine body of men it represents. The profession has soldiers, sailors, airmen, commanders and foot-soldiers. It is impacted by statesmen, politicians, writers and society, in general. What they have had to say through the course of history, on various aspects of military professionalism and conduct, is relevant to the understanding the nature of life in the military. Some quotes are over-arching and represent diverse military virtues and values. It was; therefore, best to mention them alphabetically, by authorship rather than subject.

Modern military operations place great demands on the skills and expertise of military personnel. Some demands may be unclear or open to interpretation. Practical judgment and personal moral integrity is an important facet of a military leader or commander. The military profession is about duty, but it is also about character, and the two must go together with an understanding of military values. There also needs to be knowledge of the history and principles which are the foundation of duty and military values and the role which they play in military decision making. Rules and military values are not, however, going to always provide the answers, because ultimately character is what counts.

People in the military are loyal in a way that people outside of the profession are not. Though it's a virtue in any sub-world where people act at the behest of other people, it has no other rational justification and it is not re-negotiated at every turn. It's the "plus factor" that makes it possible to count on others in a profession that can demand that one gives up his life for the sake of accomplishing a task.

There are some quotes mentioned, which are of a generic nature, or relating to other fields of human endeavour, but they have been included wherever it was felt that they have some military application. These quotes also hope to provide a basis for understanding military virtues. They help to explain where one's true duty lies and what military values actually stand for. Virtues are directed to an end and are understood more clearly when the true end of the mission, to ultimately to bring about peace and stability, is seen to be the final goal.

Duty should be enough, one might think, in any profession, but duties can conflict and the 'perfect' duty may not be recognised. Many crimes have been committed under the plea, 'I was only doing my duty'. This weakness in duty indicates that there is a need to provide ethical guidelines. History provides a platform for this. The military does this by inculcating values, based on the ancient military virtues of honour and chivalry. These values are essential to the 'moral dimension' of the profession.

Some military values are moral imperatives, such as loyalty, courage and honesty. They are not a choice. A soldier cannot choose whether to be courageous or not. Values are required in performance of duty and are expected to be conformed to. Virtue is distinguished from bravery, and still more from enthusiasm for the business of war. The first is a necessary constituent of it, but in the same way as bravery, which is a natural gift in some men, may arise in a soldier from habit and custom. It also subjugates any impulse to unbridled activity and exercise of force, and submits itself to demands of a higher kind, to obedience, order, rule, and method. Enthusiasm for the profession gives life and greater impetus to the military virtue of an army.

War is a special business, different and separate from the other pursuits which occupy the life of man. To be imbued with a sense of the military spirit requires an individual to gain confidence and expertness in it, to be completely given up to it, and to blend the military virtue of the profession of arms in the individual. Nevertheless, however much pains may be taken to combine the soldier and the citizen in one and the same individual and however much we may imagine times have changed since the ancient days, it is not possible to do away with individuality in the military. This is what makes the profession unique.

Even with the most decided inclination to look at war from the perspective of a pacifist, it would be sacrilegious to look down upon the esprit de corps which exists amongst individuals who bear arms for their country. A military which preserves itself under the heaviest fire, which is never shaken by imaginary fears, and in the face of real danger disputes the ground inch by inch is justifiably proud of its victories. It never loses its sense of obedience, its respect for and confidence in its leaders, is inured to privations and fatigue, and looks upon all its toils as the means to victory. This intrepidity is the true military spirit.

Soldiers may fight bravely and do great things without displaying this military virtue. Victories have also been attained by the use of stratagems, new weapons and equipment, and in their imaginative employment.

Military virtue is for the parts, what the genius of the commander is for the whole. The general can only guide the whole, not each separate part, and where he cannot guide the part; there military virtue must hold sway. A general is chosen by the reputation of his superior talents, after careful probation; but this probation diminishes as we descend the scale of rank, where what is wanting in talent military virtue supplies. The natural qualities of people: bravery, aptitude, powers of endurance and enthusiasm play their part. The military virtue of the profession of arms is thus one of the most important moral powers in war.

How much this refining of ore into polished metal, has been done, we see in the history of success in conflicts. When men could not reduce strategy to a formula, detailed planning necessarily failed, due to the inevitable frictions encountered: chance events, imperfections in execution, and the independent will of the opposition. Instead, the human elements were paramount: leadership, morale, and the almost instinctive savvy of the best generals. All successful military leaders did not expect a plan of operations to survive beyond the first contact with the enemy. They set only the broadest of objectives and emphasised seizing unforeseen opportunities as they arose. Strategy is not a lengthy action plan. It is the evolution of a central idea through continually changing circumstances.

Most men merely act on instinct, and the amount of success they achieve depends on the amount of talent they were born with. All great commanders have acted on instinct, and the fact that their instinct was always sound is partly the measure of their innate greatness and genius. Yet when it is not a question of acting oneself but of persuading others, the need is for clear ideas and the ability to show their connection with each other. Character thus gains a greater threshold. Quotes on these matters do, therefore, have some practical value.

I trust that the reader will enjoy reading them as much as the pleasure I had in writing them. There will always be additions and a perspicacious reader will always feel that some other quote was more relevant, apt or suitable. Suggestions or comments could be sent to me at *colharjeetsingh@yahoo.com*

CONTENTS

History is not a burden on the memory but an illumination of the soul.

- *Lord Acton*

✧✧✧

And remember, where you have a concentration of power in a few hands, all too frequently men with the mentality of gangsters get control. History has proven that.

- *Lord Acton*

✧✧✧

I cannot accept your canon that we are to judge Pope and King unlike other men with a favourable presumption that they did no wrong. If there is any presumption, it is the other way, against the holders of power, increasing as the power increases. Historic responsibility has to make up for the want of legal responsibility. All power tends to corrupt and absolute power corrupts absolutely. Great men are almost always bad men, even when they exercise influence and not authority: still more when you superadd the tendency or certainty of corruption by full authority. There is no worse heresy than that the office sanctifies the holder of it.

- *Lord Acton*

(Letter to Bishop Mandell Creighton, 1887)

✧✧✧

The long-term versus the short-term argument is one used by losers.

- *Lord Acton*

✧✧✧

Liberty is not a means to a higher political end. It is itself the highest political end. Liberty is not the power of doing what we like, but the right of being able to do what we ought.

- *Lord Acton*

(Essays on Freedom and Power)

✧✧✧

The danger is not that a particular class is unfit to govern: every class is unfit to govern.

- *Lord Acton*

✧✧✧

A wise person does at once, what a fool does at last. Both do the same thing; only at different times.

- Lord Acton

✧✧✧

Strength lies not in defence but in attack.

- Marquis de Acerba

✧✧✧

A teacher affects eternity; he can never tell, where his influence stops.

- Henry B. Adams

✧✧✧

They know enough who know how to learn.

- Henry B. Adams

✧✧✧

Human beings, who are almost in unique in having the ability to learn from the experience of others, are also remarkable for their apparent disinclination to do so.

- Douglas Adams

✧✧✧

Politics as a practice, whatever its professions, has always been a systematic organization of hatred.

- Henry Brooke Adams

✧✧✧

The declaration that our people are hostile to a government made by themselves, and for themselves, and conducted by themselves, is an insult.

- John Adams

✧✧✧

In most forms of unconventional war, the objective is the allegiance of the people around whom, and presumably on whose behalf, the conflict is taking place. The occupation of some specific bit of ground is secondary, if it matters at all.

- Thomas K. Adams
(U.S. Special Operations Forces)

✧✧✧

He wore a uniform rather as a priest wears vestments. Duty was his ever-present thought. He could forgive his men for relaxing their sense of duty when not under fire—drinking and a little quiet looting and attempted rape—but an officer and gentleman in uniform was, in his opinion, never off duty.

- Anthony Adamson, quoted in Sandra Gwyn, Tapestry of War;
A Private View of Canadians in the Great War, 1992

✧✧✧

Battles are sometimes won by generals; wars are nearly always won by sergeants and privates.

- F.E. Adcock, British classical scholar

✧✧✧

In war, truth is the first casualty.

- Aeschylus

✧✧✧

The smaller the mind the greater the conceit.

- Aesop (620–560 BC)

✧✧✧

When I think of the enlisted force I see dedication, determination, loyalty and valour.

- Chief Master Sergeant of the Air Force Paul W. Airey
First Chief Master Sergeant of the United States Air Force

✧✧✧

It were better to be a soldier's widow than a coward's wife.

- Thomas B. Aldrich, 1836-1907

✧✧✧

I do not fear an army of lions, if they are led by a lamb.
I do fear an army of sheep, if they are led by a lion.

- Alexander the Great (356 BC–323 BC)

✧✧✧

Remember upon the conduct of each depends the fate of all.

- Alexander the Great

✧✧✧

A tomb now suffices him for whom the whole world was not sufficient.

- Alexander the Great

✧✧✧

I am indebted to my father for living, but to my teacher for living well.

- Alexander the Great

✧✧✧

I am dying from the treatment of too many physicians.

- Alexander the Great

✧✧✧

Learn to obey before you command. Often the test of courage is not to die but to live.

- Vittorio Alfleri

✧✧✧

No one knows what to say in the loser's locker room.

- Muhammad Ali

✧✧✧

They knew not the day or hour nor the manner of their passing when far from home they were called to join that great band of heroic airmen that went before.

- Inscription from the American Cemetery and Memorial Cambridge England

✧✧✧

It takes great courage to faithfully follow what we know is true.

- Sara E Anderson

✧✧✧

History is replete with examples of empires mounting impressive military campaigns on the cusp of their impending economic collapse.

- Eric Alterman

✧✧✧

The Navy has both a tradition and a future—and we look with pride and confidence in both directions.

- Admiral George Anderson, 1 August 1961

✧✧✧

Innovation is the bridge between doctrine and war.

- Maj William A. Andrews, USAF

✧✧✧

Depending upon wealth, geography, and power, we perceive different threats as the most pressing. But the truth is that we cannot afford to choose. Collective security today depends on accepting that the threats which each region of the world perceives as the most urgent are in fact equally so for all. In our globalized world, the threats we face are interconnected .... [and] whatever threatens one threatens all.

- Kofi Annan
UN Secretary General

✧✧✧

To live is to choose. But to choose well, you must know who you are and what you stand for, where you want to go and why you want to get there.

- Kofi Annan

✧✧✧

It has been said that arguing against globalization is like arguing against the laws of gravity.

- Kofi Annan

✧✧✧

We may have different religions, different languages, different colour skin, but we all belong to one human race.

- Kofi Annan

✧✧✧

We have the means and the capacity to deal with our problems, if only we can find the political will.

- Kofi Annan

✧✧✧

The Lord had the wonderful advantage of being able to work alone.

- Kofi Annan

✧✧✧

You can shoot down every MiG the Soviets employ, but if you return to base and the lead Soviet tank commander is eating breakfast in your snack bar, Jack, you've lost the war.

- Anonymous A-10 Pilot, USAF

✧✧✧

We, the willing, led by the unknowing, are doing the impossible for the ungrateful. We have now done so much for so long with so little, we are now capable of doing anything with nothing.

- Anonymous

✧✧✧

Staffs analyze; commanders synthesize.

- Anonymous

✧✧✧

Remember as you follow where you may be led to regard discipline and vigilance as of first importance, and to obey with alacrity the orders transmitted to you; as nothing contributes so much to the credit and safety of an army as the union of large bodies by a single discipline.

- Archidamus, King of Sparta, exhorting the Spartans & allies at the start of the Peloponnesian War

✧✧✧

Give me a place to stand and a lever long enough and I will move the world.

- Archimedes

✧✧✧

Human war has been the most successful of our cultural traditions.

- Robert Ardrey

✧✧✧

We make war that we may live in peace.

- Aristotle

✧✧✧

Excellence is an art won by training and habituation. We do not act rightly because we have virtue or excellence, but we rather have those because we have acted rightly. We are what we repeatedly do. Excellence then, is not an act, but a habit.

- Aristotle

✧✧✧

I count him braver who overcomes his desires than him who conquers his enemies; for the hardest victory is over self.

- Aristotle

✧✧✧

Anybody can become angry that is easy, but to be angry with the right person and to the right degree and at the right time and for the right purpose, and in the right way—that is not within everybody's power and is not easy.

- Aristotle

✧✧✧

Excellence is never an accident. It is always the result of high intention, sincere effort, and intelligent execution; it represents the wise choice of many alternatives—choice, not chance, determines your destiny.

- Aristotle

✧✧✧

The whole is greater than the sum of its parts.

- Aristotle

✧✧✧

It is not enough to win a war; it is more important to organize the peace.

- Aristotle

✧✧✧

A tyrant must put on the appearance of uncommon devotion to religion. Subjects are less apprehensive of illegal treatment from a ruler whom they consider god-fearing and pious. On the other hand, they do less easily move against him, believing that he has the gods on his side.

- Aristotle

✧✧✧

For man, when perfected, is the best of animals, but, when separated from law and justice, he is the worst of all; since armed injustice is the more dangerous, and he is equipped at birth with the arms of intelligence and with moral qualities which he may use for the worst ends. Wherefore, if he have not virtue, he is the most unholy and the most savage of animals, and the most full of lust and gluttony. But justice is the bond of men in states, and the administration of justice, which is the determination of what is just, is the principle of order in political society.

- Aristotle

✧✧✧

The war we are fighting today against terrorism is a multifaceted fight. We have to use every tool in our toolkit to wage this war—diplomacy, finance, intelligence, law enforcement, and of course, military power—and we are developing new tools as we go along.

- Richard Armitage

✧✧✧

If war breaks out, our country will not be any more located in Berlin or Brandenburg, neither here nor there, but in the souls of men.

- General von Arnim

✧✧✧

Offense is the essence of air power.

- General H. H. 'Hap' Arnold, US Air Force.

✧✧✧

Strategic air assault is wasted if it is dissipated piecemeal in sporadic attacks between which the enemy has an opportunity to readjust defences or recuperate.

- General H. H. 'Hap' Arnold, US Air Force.

✧✧✧

A modern, autonomous, and thoroughly trained Air Force in being at all times will not alone be sufficient, but without it there can be no national security.

- General H. H. 'Hap' Arnold, US Air Force.

✧✧✧

As a nation we were not prepared for World War II. Yes, we won the war, but at a terrific cost in lives, human suffering, and material, and at times the margin was narrow History alone can reveal how many turning points there were, how many times we were near losing, and how our enemies' mistakes pulled us through. In the flush of victory, some like to forget these unpalatable truths.

- General H.H. "Hap" Arnold

✧✧✧

Strategic air attack is wasted if it is dissipated piecemeal in sporadic attacks between which enemy has an opportunity to readjust defences or recuperate.

- General H.H. "Hap" Arnold

✧✧✧

Air Force belongs to those who come from ranks of labour, management, the farms, the stores, the professions and colleges and legislative halls...Air Power will always be the business of every American citizen.

- General H.H. "Hap" Arnold

✧✧✧

To commit troops to a campaign in which they cannot be provided with adequate air support is to court disaster.

- Field Marshal Claude Auchinleck, 1940

✧✧✧

Never let the future disturb you. You will meet it, if you have to, with the same weapons of reason which today arm you against the present.

- Marcus Aurelius (121 AD–180 AD), Meditations.

✧✧✧

Time is like a river made up of the events which happen and its current is strong; no sooner does anything appear than it is swept away, and another comes in its place, and will be swept away too.

- Marcus Aurelius, Meditations

✧✧✧

Waste no time arguing what a good man should be. Be one.

- Marcus Aurelius

✧✧✧

A hungry dog hunts best. A hungrier dog hunts even better.

- Norman Augustine

✧✧✧

All too many consultants, when asked, 'What is 2 and 2?' respond, 'What do you have in mind?'

- Norman Augustine, Under Secretary of the US Army (1975-77)

✧✧✧

Acronyms and abbreviations should be used to the maximum extent possible to make trivial ideas profound .... Q.E.D.

- Norman Augustine

✧✧✧

Bulls do not win bull fights. People do.

- Norman Augustine

✧✧✧

If you can afford to advertise, you don't need to.

- Norman Augustine

✧✧✧

In the year 2054, the entire defence budget will purchase just one aircraft. This aircraft will have to be shared by the Air Force and Navy 3-1/2 days each per week except for leap year, when it will be made available to the Marines for the extra day.

- Norman Augustine

✧✧✧

The purpose of all war is peace.

- Saint Augustine (354-430)

✧✧✧

Festina lente: Make haste slowly.

- Emperor Augustus

✧✧✧

Batmen are officers' orderlies or body servants, and, as the officer is supposed to be the brain of the Army, it is easy to see how important a position these seemingly unimportant people hold in the military world.

Blame in the Army is very much like the rings that are caused when a small boy throws a stone into a pond. Starting from the centre they spread outward and outward until they have covered the whole face of the pond. So it is with military life. Let the general's boots not be properly cleaned, or his shaving water not the right temperature, and the neglect or carelessness of his batman has its effect in every portion of the command. Starting off in the morning in an angry mood, the general strafes the battalion commanders, they pass it on with additions to the company commanders, who in their turn add a little and hand it on to the platoon sergeants.

The sergeants beautify the language and hurl it at the corporals, and the corporals, having flavoured it still a little more, throw it at the privates, who look as if they were taking it all in, while all the time they are wondering when the parade will be dismissed and they can get away to things that have more appeal for them. On the other hand, let the general's shaving water be of the right temperature, or his boots shining and in a high state of polish, and everything is quiet and peaceful.

The sailor and soldier are unique in one respect, in that they are brought up on theory, and, only perhaps once in twenty years, get the opportunity to put these theories into practice. When this does happen, they usually find that the maxims and precepts they have so carefully committed to memory do not meet the case.

- John Aye, Humour in the Army, 1932

✧✧✧

B

Courage, however, is that firmness of spirit, that moral backbone, which, fully appreciating the danger involved, nevertheless goes on with the undertaking. Bravery is physical; courage is mental and moral. You can be cold all over; your hands may tremble; your legs may quake; your knees be ready to give way—that is fear. If, nevertheless, you go forward; if in spite of this physical defection you continue to lead your men against the enemy, you have the courage. The physical manifestation of fear will pass away. You may experience them but once.

- Major C. A. Bach, US Army Instructor

✧✧✧

He that commands the sea, is at great liberty, and may take as much, and as little, of the war as he will.

- Francis Bacon
(English philosopher, statesman, scientist, lawyer, jurist, author)
Lord High Chancellor (1617–1621)

✧✧✧

Any moral action is the action of the human will, which is governed by belief and spurred on by the passions.

- Francis Bacon

✧✧✧

As for the philosophers, they make imaginary laws for imaginary commonwealths; and their discourses are as the stars, which give little light, because they are so high.

- Francis Bacon (On the Advancement of Learning)

✧✧✧

Men have sought to make a world from their own conception and to draw from their own minds all the material which they employed, but if, instead of doing so, they had consulted experience and observation, they would have the facts and not opinions to reason about, and might have ultimately arrived at the knowledge of the laws which govern the material world.

- Francis Bacon

✧✧✧

No universal rules can be made, as both situations and men's characters differ.

- Francis Bacon

✧✧✧

Good habit is what aids men in directing their will toward the good.

- Francis Bacon

✧✧✧

Printing, gunpowder and the compass: These three have changed the whole face and state of things throughout the world; the first in literature, the second in warfare, the third in navigation; whence have followed innumerable changes, in so much that no empire, no sect, no star seems to have exerted greater power and influence in human affairs than these mechanical discoveries.

- Francis Bacon

✧✧✧

The successful pilots succeeded because they did not open fire until they were close to the target.

- Group Captain Douglas Bader, CBE, DSO & Bar, DFC & Bar

✧✧✧

If you had the height, you controlled the battle.
If you came out of the sun, the enemy could not see you.
If you held your fire until you were very close, you seldom missed.

- Group Captain Douglas Bader, CBE, DSO & Bar, DFC & Bar

✧✧✧

The core of a soldier is moral discipline. It is intertwined with the discipline of physical and mental achievement. Total discipline overcomes adversity, and physical stamina draws on an inner strength that says "drive on."

- William G. Bainbridge, Sergeant Major of the Army

✧✧✧

If you're not gonna pull the trigger, don't point the gun.

- James Baker, US Secretary of State (1989–1992)

✧✧✧

That idea is so damned nonsensical and impossible that I'm willing to stand on the bridge of a battleship while that nitwit tries to hit if from the air.

- Newton D. Baker, U.S. Secretary of War,
regarding the idea of airplanes sinking a battleship.

✧✧✧

War would end if the dead could return.

- Stanley Baldwin, statesman (1867-1947)

✧✧✧

Let us never forget this: since the day of the air, the old frontiers are gone. When you think of the defence of England you no longer think of the chalk cliffs of Dover; you think of the Rhine. That is where our frontier lies.

- Stanley Baldwin, 1934

✧✧✧

The bomber will always get through. The only defence is in offence, which means that you have to kill more women and children more quickly than the enemy if you want to save yourselves.

- Stanley Baldwin, British Prime Minister, House of Commons speech, 10 November 1932

✧✧✧

Since the day of the air, the old frontiers are gone. When you think of the defence of England you no longer think of the chalk cliffs of Dover; you think of the Rhine. That is where our frontier lies.

- Stanley Baldwin, House of Commons speech, 30 July 1934

✧✧✧

I wish for many reasons flying had never been invented.

- Stanley Baldwin, on learning that Germany had secretly built an air force in defiance of the Treaty of Versailles, 1935.

✧✧✧

What the proprietorship of these papers is aiming at is power, and power without responsibility is the prerogative of the harlot through the ages.

- Stanley Baldwin (1867-1947)

✧✧✧

You do not have to blow out the other fellow's light to let your own shine.

- Bernard Baruch

✧✧✧

Peace can be made tranquil and secure only by understanding and agreement fortified by sanctions. We must embrace international cooperation or international disintegration.

- Bernard Baruch

✧✧✧

Be who you are and say what you feel, because those who mind don't matter and those who matter don't mind.

- Bernard Baruch

✧✧✧

I'm not smart. I try to observe. Millions saw the apple fall but Newton was the one who asked why.

- Bernard Baruch

✧✧✧

Terrorism, like viruses, is everywhere. There is a global perfusion of terrorism, which accompanies any system of domination as though it were its shadow, ready to activate itself anywhere, like a double agent.

- Jean Baudrillard, The Spirit of Terrorism and Other Essays

✧✧✧

Among the subalterns [of the 2nd Scottish rifles] is Somervail, who is to be the lone survivor among the officers at the end of the battle [of Neuve Chapelle], and is to bring out the unbroken remnant of the battalion in five nights' time. Finally the long column passes, and another Regiment looms out of the darkness, for the road from Estaires is full of troops tonight. As the last man passes from sight, and the picture fades back into its proper place in the past, one wants to know on what foundations the fortitude of this battalion was built. can one pick out the main ingredients of its high morale? Out of the factors I have brought into this study I consider five to be the most important, which I have arranged in order of priority.

First, I would place Regimental loyalty; the pride in belonging to a good battalion, in knowing other people well and being known by them; in having strong roots in a well-loved community.

Second, the excellent officer-other rank relationship; the high quality of the leaders, and the trust placed in them by their men; the mutual confidence and goodwill which developed in the harsh life of the trenches.

Third, strong discipline; the balance between self-discipline and the imposed sort.

Fourth, the sense of duty of all ranks; highly developed in the officer by his training and background; developed in the soldier both by his training and by the realization that someone else would have to do his job if he failed to do it properly himself.

Fifth, sound administration, so that in spite of many difficulties the battalion was well provided with the necessities of war such as rations and ammunition.

- John Baynes, "The Regimental System"

✧✧✧

The difficulties of peace are better than the agony of war.

- Israeli Prime Minister Menachem Begin

✧✧✧

Things you see from there are not what you see from here.

- Israeli Prime Minister Menachem Begin

✧✧✧

Trying to use the lessons of the past correctly poses two dilemmas. One is the problem of balance: knowing how much to rely on the past as a guide and how much to ignore it. The other is the problem of selection: certain lessons drawn from experience contradict others.

- Richard Betts

Soldiers, Statesmen, and Cold War Crises

✧✧✧

Coward: one who, in a perilous emergency, thinks with his legs.

- Ambrose Bierce, American writer (1842 –1913)

✧✧✧

War is God's way of teaching Americans geography.

- Ambrose Bierce

✧✧✧

If you think education is expensive, try ignorance.

- Derek Bok

✧✧✧

The only thing necessary for the triumph of evil is for good men to do nothing.

- Edmund Burke

✧✧✧

No person made a greater mistake than the one who did nothing because he or she could only do a little.

- Edmund Burke

✧✧✧

For in this modern world, the instruments of warfare are not solely for waging war. Far more importantly, they are the means for controlling peace. Naval officers must therefore understand not only how to fight a war, but how to use the tremendous power which they operate to sustain a world of liberty and justice, without unleashing the powerful instruments of destruction and chaos that they have at their command.

- Admiral Arleigh Burke, 1 August 1961

✧✧✧

Throughout the entire course of history, warfare is always changing.

- General Andre Beaufre
Chief of the General Staff of the Supreme Headquarters,
Allied Powers in Europe, 1958

✧✧✧

No explanation for the current strategic situation is satisfactory without a definition of the nuclear situation; no definition of the nuclear situation is possible without knowledge of the laws that rule deterrence.

- General Andre Beaufre

✧✧✧

The game of strategy can, like music, be played in two keys. The major key is direct strategy, in which force is the essential factor. The minor key is indirect strategy, in which force recedes into the background and its place is taken by psychology and planning.

- General Andre Beaufre

✧✧✧

A General Thimayya is not born in every generation. The like of him there will seldom be—a soldier, a general, a man's man; the Army his soul, his soul, the Army.

- General PS Bhagat, VC

✧✧✧

The military wants a system that protects its policies and privileges.

- Benazir Bhutto

✧✧✧

Pakistanis will eat grass but make a nuclear bomb.

- Zulfikar Ali Bhutto (1928–1979)

✧✧✧

Pakistan was once called the most allied ally of the United States. We are now the most non-allied.

- Zulfikar Ali Bhutto, 6 July 1973

✧✧✧

Religion is a link between God and man and man and man. Political ideology is a link between man and man. For this reason the great religions of the world like Hinduism, Buddhism, Judaism, Christianity and Islam, the last of all religions, have outlived and outlasted political ideologies. If an unlearned adventurer in his quest for political power and perpetuation brings religion down from its celestial plane to a mundane level by converting it into a narrow political ideology, the adventurer endangers the link between God and man and man and man.

- Zulfikar Ali Bhutto

✧✧✧

For if the trumpet gives an uncertain sound, who shall prepare himself to the battle?

- The Bible (Corinthians 14:8)

✧✧✧

The race is not necessarily to the swift

- The Bible (Ecclesiastes)

✧✧✧

There is no man that hath power over the spirit to retain the spirit; neither hath he power over the day of death; and there is no discharge in war: neither shall wickedness deliver him that is given to it.

- The Bible (Ecclesiastes, Chapter 8:8)

✧✧✧

And he will judge between the nations, and will decide concerning many peoples; and they shall beat their swords into ploughshares, and their spears into pruning-hooks; nation shall not lift up sword against nation, neither shall they learn war any more.

- The Bible (Isaiah, Chapter 2:4)

✧✧✧

When a strong man, fully armed, guards his palace, his possessions are safe.

- The Bible (Luke 11:21)

✧✧✧

Every purpose is established by counsel; And by wise guidance make thou war.

- The Bible (Proverbs, Chapter 20:18)

✧✧✧

The most powerful military in the world cannot invade, kill or capture a network or destroy every loose weapon on the planet. The best response to this network of terror is to build a network of our own—a network of like-minded countries and organizations that pools resources, information, ideas, and power. Taking on the radical fundamentalists alone isn't necessary, it isn't smart, and it won't succeed.

- Joe Biden, 7 August 2006

✧✧✧

Such a man may not distinguish himself so much in the eyes of others but he will have sacrificed more and suffered greater demands than his more outwardly courageous companions. This is why I have always respected the courage of 'the hundred others' without whom no commander or leader can fight his ship, his aircraft, or his army divisions effectively.

- General Sir Peter de la Billière, commander of British forces in the First Iraq War gave all his men credit when he recognized that not all men can act courageously as easily as others; some have to push themselves hard to carry out their duties.

✧✧✧

Fools say that they learn by experience. I prefer to profit by others' experience.

- Otto Von Bismarck (German Chancellor, 1871-1890)

✧✧✧

Think with your blood.

- Otto von Bismarck

✧✧✧

Anyone who has ever looked into the glazed eyes of a soldier dying on the battlefield will think hard before starting a war.

- Otto von Bismarck

✧✧✧

People never lie so much as after a hunt, during a war or before an election.

- Otto Von Bismarck

✧✧✧

Politics is not an exact science.

- Otto Von Bismarck

✧✧✧

No bird soars too high if he soars with his own wings.

- William Blake

✧✧✧

There is a difference between interest and commitment. When you are interested in doing something, you do it only when it is convenient. When you are committed to something, you accept no excuses, only results.

- Kenneth Blanchard

✧✧✧

I found it peculiar that those who wanted to take military action could—with 100 per cent certainty—know that the weapons existed and turn out to have zero knowledge of where they were.

- Hans Blix

✧✧✧

Ever forward, but slowly.

- Gebhard Lebrecht von Blucher

Prussian Field Marshal who led his army against Napoleon I at the Battle of the Nations at Leipzig in 1813 and at the Battle of Waterloo in 1815 with the Duke of Wellington

✧✧✧

Raise high the black flags, my children. No prisoners. No pity. I will shoot any man I see with pity in him.

- Gebhard Lebrecht von Blucher

✧✧✧

A soldier is he whose blood makes the glory of the general.

- Henry G. Bohn

✧✧✧

Every dog is a lion at home.

- Henry G. Bohn

✧✧✧

He who knows himself best esteems himself least.

- Henry G. Bohn

✧✧✧

One of these days is none of these days.

- Henry G. Bohn

✧✧✧

Prediction is very difficult, especially about the future.

- Niels Bohr (1885-1962)

Danish physicist who made fundamental contributions to understanding atomic structure & quantum mechanics, for which he received the Nobel Prize in Physics in 1922.

✧✧✧

An expert is someone who knows some of the worst mistakes, which can be made, in a very narrow field.

- Niels Bohr (1885-1962)

✧✧✧

Every great and deep difficulty bears in itself its own solution. It forces us to change our thinking in order to find it.

- Niels Bohr (1885-1962)

✧✧✧

Never express yourself more clearly than you are able to think.

- Niels Bohr (1885-1962)

✧✧✧

One individual may die for an idea; but that idea will, after his death, incarnate itself in a thousand lives. That is how the wheel of evolution moves on and the ideas and dreams of one nation are bequeathed to the next.

- Netaji Subhash Chandra Bose

✧✧✧

Freedom is not given, it is taken.

- Netaji Subhash Chandra Bose

✧✧✧

Machines don't fight wars. People do, and they use their minds.

- Col John R. Boyd, US Air Force

✧✧✧

People, ideas, and hardware—In that order!

- Col John R. Boyd, US Air Force

✧✧✧

Complexity (technical, organizational, operational, etc.) causes commanders and subordinates alike to be captured by their own internal dynamics or interaction; hence they cannot adapt to rapidly changing external (or even internal) circumstances.

- Col John R. Boyd, US Air Force

✧✧✧

Harmony in operations is created by the bonds of implicit communications and trust that evolve as a consequence of the similar mental images or impressions each individual creates and commits to memory by repeatedly sharing the same variety of experience in the same ways.

- Col John R. Boyd, US Air Force

✧✧✧

Leadership is the art of inspiring people to enthusiastically take action toward the achievement of uncommon goals.

- Col John R. Boyd, US Air Force

✧✧✧

Leadership is the art of inspiring people to enthusiastically take action toward the achievement of uncommon goals.

- Col John R. Boyd, US Air Force

✧✧✧

[Strategy is] a mental tapestry of changing intentions for harmonizing and focusing our efforts as a basis for realizing some aim or purpose in an unfolding and often unforeseen world of many bewildering events and many contending interests. [Its aim was] to improve our ability to shape and adapt to unfolding circumstances, so that we (as individuals or as groups or as a culture or as a nation-state) can survive on our own terms.

- Col John R. Boyd, US Air Force

✧✧✧

Every plan of campaign ought to have several branches and to have been so well thought out that one or other of the said branches cannot fail of success.

- Bourchet

✧✧✧

Soldiers usually win the battles and generals get the credit for them.

- Napoleon Bonaparte

✧✧✧

There are two means of moving men—interest and fear.

- Napoleon Bonaparte

✧✧✧

God gave men dominion over the beasts and not over his fellow men unless they submit of their own free will.

- Napoleon Bonaparte

✧✧✧

In war, the moral is to the physical as three to one.

- Napoleon Bonaparte

✧✧✧

There are times when a battle decides everything, and there are times when the most insignificant thing can decide the outcome of a battle.

- Napoleon Bonaparte

✧✧✧

There are but two powers in the world, the sword and the mind. In the long run the sword is always beaten by the mind.

- Napoleon Bonaparte

✧✧✧

The word impossible is not in my dictionary.

- Napoleon Bonaparte

✧✧✧

The most dangerous moment comes with victory.

- Napoleon Bonaparte

✧✧✧

The first virtue in a soldier is endurance of fatigue; courage is only the second virtue.

- Napoleon Bonaparte

✧✧✧

Strategy is the art of making use of time and space. I am less concerned about the later than the former. Space we can recover, lost time never.

- Napoleon Bonaparte

✧✧✧

The first shot is for the Devil, the second for God, and only the third for the King.

- Napoleon Bonaparte

✧✧✧

Nothing is more destructive than the charge of Artillery on a crowd.

- Napoleon Bonaparte

✧✧✧

The best generals are those who have served in the Artillery.

- Napoleon Bonaparte

✧✧✧

God fights on the side with the best Artillery.
With Artillery, War is made.

- Napoleon Bonaparte

✧✧✧

Leave the Artillerymen alone, they are an obstinate lot.

- Napoleon Bonaparte

✧✧✧

In war nothing is achieved except by calculation. Everything that is not soundly planned in its details yields no results.

- Napoleon Bonaparte

✧✧✧

To dogmatize upon that which you have not practised is the prerogative of ignorance.

- Napoleon Bonaparte

✧✧✧

It may be that in future I may lose a battle, but I shall never lose a minute.

- Napoleon Bonaparte

✧✧✧

Read over and over again the campaigns of Alexander, Hannibal, Caesar, Gustavus, Turenne, Eugene and Frederic. ... This is the only way to become a great general and master the secrets of the art of war.

- Napoleon Bonaparte

✧✧✧

A soldier will fight long and hard for a bit of coloured ribbon.

- Napoleon Bonaparte

✧✧✧

How many things apparently impossible have nevertheless been performed by resolute men who had no alternative but death.

- Napoleon Bonaparte

✧✧✧

Two armies are two bodies which meet and try to frighten each other.

- Napoleon Bonaparte

✧✧✧

The torment of precautions often exceeds the dangers to be avoided. It is sometimes better to abandon one's self to destiny.

- Napoleon Bonaparte

✧✧✧

I love a brave soldier who has undergone the baptism of fire.

- Napoleon Bonaparte

✧✧✧

The secret of war lies in the communications.

- Napoleon Bonaparte

✧✧✧

Every soldier carries a marshal's baton in his pack.

- Napoleon Bonaparte

✧✧✧

If they want peace, nations should avoid the pin-pricks that precede cannonshots.

- Napoleon Bonaparte

✧✧✧

An army marches on its stomach.

- Napoleon Bonaparte

✧✧✧

You must not fight too often with one enemy, or you will teach him all your art of war.

- Napoleon Bonaparte

✧✧✧

In war there is but one favourable moment; the great art is to seize it!

- Napoleon Bonaparte

✧✧✧

One bad general is worth two good ones.

- Napoleon Bonaparte

✧✧✧

There are certain things in war of which the commander alone comprehends the importance. Nothing but his superior firmness and ability can subdue and surmount all difficulties.

- Napoleon Bonaparte

✧✧✧

He that makes war without many mistakes has not made war very long.

- Napoleon Bonaparte

✧✧✧

The most important qualification of a soldier is fortitude under fatigue and privation. Courage is only second; hardship, poverty and want are the best school for a soldier.

- Napoleon Bonaparte

✧✧✧

It should not be believed that a march of three or four days in the wrong direction can be corrected by a countermarch. As a rule, this is to make two mistakes instead of one.

- Napoleon Bonaparte

✧✧✧

In time of revolution, with perseverance and courage, a soldier should think nothing impossible.

- Napoleon Bonaparte

✧✧✧

An army's effectiveness depends on its size, training, experience, and morale, and morale is worth more than any of the other factors combined.

- Napoleon Bonaparte

✧✧✧

Between a battle lost and a battle won, the distance is immense and there stand empires.

- Napoleon Bonaparte

✧✧✧

In war, three-quarters turns on personal character and relations; the balance of manpower and materials counts only for the remaining quarter.

- Napoleon Bonaparte

✧✧✧

It would be a joke if the conduct of the victor had to be justified to the vanquished.

- Napoleon Bonaparte

✧✧✧

In war, as in politics, no evil—even if it is permissible under the rules—is excusable unless it is absolutely necessary. Everything beyond that is a crime.

- Napoleon Bonaparte

✧✧✧

If you wage war, do it energetically and with severity. This is the only way to make it shorter and consequently less inhuman.

- Napoleon Bonaparte

✧✧✧

There are in Europe many good generals, but they see too many things at once. I see one thing, namely the enemy's main body. I try to crush it, confident that secondary matters will then settle themselves.

- Napoleon Bonaparte

✧✧✧

There is no man more pusillanimous than I when I am planning a campaign. I purposely exaggerate all the dangers and all the calamities that the circumstances make possible. I am in a thoroughly painful state of agitation. This does not keep me from looking quite serene in front of my entourage; I am like an unmarried girl labouring with child. Once I have made up my mind, everything is forgotten except what leads to success.

- Napoleon Bonaparte

✧✧✧

Audacity succeeds as often as it fails; in life it has an even chance.

- Napoleon Bonaparte

✧✧✧

If you had seen one day of war, you would pray to God that you would never see another.

- Napoleon Bonaparte

✧✧✧

The basic principle that we must follow in directing the armies of the Republic is this: that they must feed themselves on war at the expense of the enemy territory.

- Napoleon Bonaparte

✧✧✧

Sometimes a single battle decides everything and sometimes, too, the slightest circumstance decides the issue of a battle. There is a moment in every battle at which the least manoeuvre is decisive and gives superiority, as one drop of water causes overflow.

- Napoleon Bonaparte

✧✧✧

You do not get peace by shouting: Peace. Peace is a meaningless word; what we need is a glorious peace.

- Napoleon Bonaparte

✧✧✧

If the art of war were nothing but the art of avoiding risks, glory would become the prey of mediocre minds. I have made all the calculations, fate will do the rest.

- Napoleon Bonaparte

✧✧✧

An Emperor confides in national soldiers, not in mercenaries.

- Napoleon Bonaparte

✧✧✧

The fate of a Nation may sometimes depend upon the position of a fortress.

- Napoleon Bonaparte

✧✧✧

Men soon get tired of shedding their blood for the advantage of a few individuals, who think they amply reward the soldiers' perils with the treasures they amass.

- Napoleon Bonaparte

✧✧✧

It is the business of cavalry to follow up the victory, and to prevent the beaten army from rallying.

- Napoleon Bonaparte

✧✧✧

An army which cannot be reinforced is already defeated.

- Napoleon Bonaparte

✧✧✧

The moment of greatest peril is the moment of victory.

- Napoleon Bonaparte

✧✧✧

It is by the eyes of the mind, by reasoning over the whole, by a species of inspiration that the general sees, knows and judges.

- Napoleon Bonaparte

✧✧✧

When I wish to interrupt one train of thought, I shut that drawer and open another. Do I wish to sleep? I simply close all the drawers and there I am 'asleep'.

- Napoleon Bonaparte

✧✧✧

Yes, we need forward thinkers. . . . It is also essential that we do not believe that we possess such enormous wisdom that we can dismiss the past.

- Napoleon Bonaparte

✧✧✧

How many seemingly impossible things have been accomplished by resolute men because they had to do, or die.

- Napoleon Bonaparte

✧✧✧

It is often in the audacity, in the steadfastness, of the general that the safety and the conservation of his men are found.

- Napoleon Bonaparte

✧✧✧

The military principles of Caesar were those of Hannibal, and those of Hannibal were those of Alexander—to hold his forces in hand, not to be vulnerable at any point, to throw all his forces with rapidity on any given point.

- Napoleon Bonaparte

✧✧✧

At the beginning of a campaign it is important to consider whether or not to move forward; but when one has taken the offensive it is necessary to maintain it to the last extremity.

- Napoleon Bonaparte

✧✧✧

Changing from the defensive to the offensive is one of the most delicate operations in war.

- Napoleon Bonaparte

✧✧✧

Never march by flank in front of an army in position. This principle is absolute.

- Napoleon Bonaparte

✧✧✧

In a battle, as in a siege, the art consists in concentrating very heavy fire on a particular point. The line of battle once established the one who has the ability to concentrate an unlocked for mass of artillery suddenly and unexpectedly on one of these points is sure to carry the day.

- Napoleon Bonaparte

✧✧✧

There is only one favourable moment in war; talent consists in knowing how to seize it.

- Napoleon Bonaparte

✧✧✧

In war, theory is all right so far as general principles are concerned; but in reducing general principles to practice there will always be danger. Theory and practice are the axis about which the sphere of accomplishment revolves.

- Napoleon Bonaparte

✧✧✧

The secret of great battles consists in knowing how to deploy and concentrate at the right time.

- Napoleon Bonaparte

✧✧✧

The art of war consists in being always able, even with an inferior army, to have stronger forces than the enemy at the point of attack or the point which is attacked.

- Napoleon Bonaparte

✧✧✧

The praises of enemies are always to be suspected. A man of honour will not permit himself to be flattered by them, except when they are given after the cessation of hostilities.

- Napoleon Bonaparte

✧✧✧

Gentleness, good treatment, honour the victor and dishonour the vanquished, who should remain aloof and owe nothing to pity—In war, audacity is the finest calculation of genius.

- Napoleon Bonaparte

✧✧✧

In war, groping tactics, half-way measures, lose everything.

- Napoleon Bonaparte

✧✧✧

To plan to reserve cavalry for the finish of the battle, is to have no conception of the power of combined infantry and cavalry charges, either for attack or for defence.

- Napoleon Bonaparte

✧✧✧

Victory in war does not depend entirely upon numbers or mere courage; only skill and discipline will insure it.

- Napoleon Bonaparte

✧✧✧

An army ought to only have one line of operation. This should be preserved with care, and never abandoned but in the last extremity.

- Napoleon Bonaparte

✧✧✧

Charges of cavalry are equally useful at the beginning, the middle and the end of a battle. They should be made always, if possible, on the flanks of the infantry, especially when the latter is engaged in front.

- Napoleon Bonaparte

✧✧✧

When you determine to risk a battle, reserve to yourself every possible chance of success, more particularly if you have to deal with an adversary of superior talent, for if you are beaten, even in the midst of your magazines and your communications, woe to the vanquished!

- Napoleon Bonaparte

✧✧✧

When you have resolved to fight a battle, collect your whole force. Dispense with nothing. A single battalion sometimes decides the day.

- Napoleon Bonaparte

✧✧✧

The transition from the defensive to the offensive is one of the most delicate operations in war.

- Napoleon Bonaparte

✧✧✧

The first qualification of a soldier is fortitude under fatigue and privation. Courage is only the second; hardship, poverty, and want are the best school for a soldier.

- Napoleon Bonaparte

✧✧✧

In war, the general alone can judge of certain arrangements. It depends on him alone to conquer difficulties by his own superior talents and resolution.

- Napoleon Bonaparte

✧✧✧

Never lose sight of this maxim, that you should establish your cantonments at the most distant and best protected point from the enemy, especially where a surprise is possible. By this means you will have time to unite all your forces before he can attack you.

- Napoleon Bonaparte

✧✧✧

Artillery is more essential to cavalry than to infantry, because cavalry has no fire for its defence, but depends on the sabre.

- Napoleon Bonaparte

✧✧✧

Unity of command is essential to the economy of time. Warfare in the field was like a siege: by directing all one's force to a single point a breach might be made, and the equilibrium of opposition destroyed.

- Napoleon Bonaparte

✧✧✧

A general-in-chief should ask himself several times in the day, 'What if the enemy were to appear now in my front, or on my right, or my left?"

- Napoleon Bonaparte

✧✧✧

Generals who save troops for the next day are always beaten.

- Napoleon Bonaparte

✧✧✧

The worse the troops the greater the need of artillery.

- Napoleon Bonaparte

✧✧✧

We should always go before our enemies with confidence, otherwise our apparent uneasiness inspires them with greater boldness.

- Napoleon Bonaparte

✧✧✧

A man does not have himself killed for a few halfpence a day or for a petty distinction. You must speak to the soul in order to electrify the man.

- Napoleon Bonaparte

✧✧✧

In politics, stupidity is not a handicap.

- Napoleon Bonaparte

✧✧✧

A Commander-in-Chief is not exonerated from his mistakes in war, committed by virtue of an order of his Sovereign or Minister; where he that gives it is far from the field of operations, and knows little or nothing of the latest developments. Hence, it follows that any Commander-in-Chief who undertakes to execute a plan which he considers bad, is guilty. He should give his reasons, insist that the plan be changed and finally resign, rather than become the instrument of ruin of the King.

- Napoleon Bonaparte

✧✧✧

A river is an obstacle which can only delay the enemy for a few days. If you decide to defend it, the only solution is to assemble your troops as a mass and fall upon the enemy before his crossing is completed.

- Napoleon Bonaparte

✧✧✧

Nothing is more difficult, and therefore more precious, than to be able to decide.

- Napoleon Bonaparte

✧✧✧

It's pretty clear to me that as we get smaller, in fact as each of our services get smaller, we are going to need to be more efficient. That means we have to know what our requirements are and the military services must work together better so that we don't duplicate each other unnecessarily. We also have to buy things smarter.

- Admiral Jeremy Michael Boorda, US Navy

✧✧✧

The general who sets out to fight a battle by getting on with it and then adjusting may lose out to the general who has put in some thinking in the first place.

- Edward de Bono

✧✧✧

Logic never changed a belief.

- Edward de Bono

✧✧✧

An idea that is developed and put into action is more important than an idea that exists only as an idea.

- Edward de Bono

✧✧✧

Argument is meant to reveal the truth, not to create it.

- Edward de Bono

✧✧✧

If you never change your mind, why have one?

- Edward de Bono

✧✧✧

Most of the mistakes in thinking are inadequacies of perception rather than mistakes of logic.

- Edward de Bono

✧✧✧

Sometimes the situation is only a problem because it is looked at in a certain way. Looked at in another way, the right course of action may be so obvious that the problem no longer exists.

- Edward de Bono

✧✧✧

The need to be right all the time is the biggest bar to new ideas.

- Edward de Bono

✧✧✧

Only one military organization can hold and gain ground in war—a ground army supported by tactical aviation with supply lines guarded by the navy.

- General Omar N. Bradley
First Chairman of the US Joint Chiefs of Staff

✧✧✧

Man for man one division is as good as another. They vary only in the skill and leadership of their commanders.

- General Omar N. Bradley

✧✧✧

For most men the matter of learning is one of personal preference. But for army officers the obligation to learn, to grow in their profession is clearly a public duty.

- General Omar N. Bradley

✧✧✧

Amateurs talk strategy, professionals talk logistics.

- General Omar N. Bradley

✧✧✧

Leadership is intangible, and therefore no weapon ever designed can replace it.

- General Omar N. Bradley

✧✧✧

I am convinced that the best service a retired general can perform is to turn in his tongue along with his suit and to mothball his opinions.

- General Omar N. Bradley

✧✧✧

Inferior inducements bring second-rate men. Second-rate men invite second-best security. In war, there is no prize for the runner-up.

- General Omar N Bradley

✧✧✧

Airpower has become predominant, both as a deterrent to war, and—in the eventuality of war—as the devastating force to destroy an enemy's potential and fatally undermine his will to wage war.

- General Omar N. Bradley

✧✧✧

Ours is a world of nuclear giants and ethical infants. We know more about war than about peace, more about killing than we know about living.

- General Omar N. Bradley

✧✧✧

Leadership is intangible, and therefore no weapon ever designed can replace it.

- General Omar N. Bradley

✧✧✧

This is as true in everyday life as it is in battle: we are given one life and the decision is ours whether to wait for circumstances to make up our mind, or whether to act, and in acting, to live.

- General Omar N. Bradley

✧✧✧

We need to learn to set our course by the stars, not by the light of every passing ship.

- General Omar N. Bradley

✧✧✧

Only those to whom the study of war is novel permit themselves to be swept away by novel elements in the present war.

- Bernard Brodie

✧✧✧

One of the few unequivocally sound lessons of history is that the lessons we should learn are usually learned imperfectly if at all.

- Bernard Brodie

✧✧✧

If I should die, think only this of me:
That there's some corner of a foreign field
That is forever England. There shall be
In that rich earth a richer dust concealed;
A dust whom England bore, shaped, made aware,
Gave, once, her flowers to love, her ways to roam,
A body of England's, breathing English air,
Washed by the rivers, blest by suns of home.

And think, this heart, all evil shed away,
A pulse in the eternal mind, no less
Gives somewhere back the thoughts by England given;
Her sights and sounds; dreams happy as her day;
And laughter, learnt of friends; and gentleness,
In hearts at peace, under an English heaven.

- Rupert Brooke, 1914 (The Soldier)

✧✧✧

Terrorism is not an expression of rage. Terrorism is a political weapon. Remove a government's facade of infallibility, and you remove its people's faith.

- Dan Brown, Angels & Demons

✧✧✧

It is a brave act of valour to condemn death, but where life is more terrible than death it is then the truest valour to dare to live.

- Sir Thomas Brown

✧✧✧

Regret what? The secret operation was an excellent idea. It drew the Russians into the Afghan trap and you want me to regret it? On the day that the Soviets officially crossed the border, I wrote to President Carter, saying, in essence: 'We now have the opportunity of giving to the USSR its Vietnam War.'

- Zbigniew Brzezinski
US National Security Advisor (1977 -1981)

✧✧✧

It is better to conquer yourself than to win a thousand battles. Then the victory is yours. It cannot be taken from you, not by angels or by demons, heaven or hell.

- Buddha

✧✧✧

When the stakes warrant, where and when force can be effective, where no other policies are likely to be effective, when its application can be limited in scope and time, and where the potential benefits justify the potential costs and sacrifice. There can be no single or simple set of fixed rules for using force.

- President George Bush

✧✧✧

What's the sense of sending $2 million missiles to hit a $10 tent that's empty?

- President George W. Bush, Oval Office meeting, 13 September 2001.

✧✧✧

Whether we bring our enemies to justice or bring justice to our enemies, justice will be done.

- President George W. Bush

✧✧✧

We have learned that terrorist attacks are not caused by the use of strength; they are invited by the perception of weakness. And the surest way to avoid attacks on our own people is to engage the enemy where he lives and plans. We are fighting that enemy in Iraq and Afghanistan today so that we do not meet him again on our own streets, in our own cities.

- President George W. Bush

✧✧✧

We must take the battle to the enemy, disrupt his plans and confront the worst threats before they emerge.

- President George W. Bush,
U.S. Military Academy, West Point, June 2002

✧✧✧

What our enemies have begun, we will finish.

- President George W. Bush,
Address to the Nation, 9/11/02

✧✧✧

We do not create terrorism by fighting the terrorists. We invite terrorism by ignoring them.

- George W. Bush, 18 December 2005

✧✧✧

A soldier can walk the battlefields he once fought; a Marine can walk the beaches he once stormed; but an Airman can never visit the patch of sky he raced across on a mission to defend freedom. And so it's fitting that the men and women of the Air Force will have this memorial, a place here on the ground that recognizes their achievements and sacrifices in the skies above.

- President George W. Bush
Dedication of the United States Air Force Memorial , 14 October 2006

✧✧✧

Conflict behaviour is actions undertaken by one social entity in any given situation of conflict aimed at the opposing party with the intention of making that opponent abandon/modify its goals.

- Barry Buzan

✧✧✧

A conflict situation is one in which two or more social entities, however defined or structured, perceive that they possess mutually incompatible goals.

- Barry Buzan

✧✧✧

C

"Veni, Vedi, Vici"—I came, I saw, I conquered.

- Julius Caesar

✧✧✧

As a rule, men worry more about what they can't see than about what they can.

- Julius Caesar

✧✧✧

Fortune, which has a great deal of power in other matters but especially in war, can bring about great changes in a situation through very slight forces.

- Julius Caesar

✧✧✧

In war, events of importance are the result of trivial causes.

- Julius Caesar

✧✧✧

What we wish, we readily believe, and what we ourselves think, we imagine others think also.

- Julius Caesar

✧✧✧

There is a tide in the affairs of men
Which taken at the flood, leads on to fortune;
Omitted, all the voyage of their life
Is bound in shallows and in miseries.

- Julius Caesar
(from the play by William Shakespeare)

✧✧✧

Iacta alea est, inquit.

The die is cast.

- Julius Caesar

Said when crossing the river Rubicon with his legions on 10 January, 49 BC, thus beginning the civil war with the forces of Pompey. The Rubicon river was the boundary of Gaul, the province Caesar had the authority to keep his army in. By crossing the river, he had committed an invasion of Italy.

✧✧✧

[The good diplomat] will never rely for the success of his mission either on bad faith or on promises that he cannot execute. It is a fundamental error, and one widely held, that a clever negotiator must be a master of deceit. Deceit is indeed the measure of smallness of mind of him who uses it; it proves that he does not possess sufficient intelligence to achieve results by just and reasonable means. Honesty is here and everywhere the best policy ... Apart from the fact that a lie is unworthy of a great Ambassador, it actually does more harm than good to negotiation, since though it may confer success today, it will create an atmosphere of suspicion which tomorrow will make further success impossible ... The negotiator therefore must be a man of probity and one who loves truth; otherwise he will fail to inspire confidence.

- Francois de Callieres
On the Manner of Negotiating with Princes, 1716

✧✧✧

[Diplomacy] is a profession by itself which deserves the same profession and assiduity of attention that men give to other recognized professions. ... There are many qualities which may be developed with practice, and the greatest part of the necessary knowledge can only be acquired, by constant application to the subject. In this sense, diplomacy is certainly a profession itself capable of occupying a man's whole career, and those who think to embark upon a diplomatic mission as a pleasant diversion from their common task only prepare disappointment for themselves and disaster for the cause which they serve.

- Francois de Callieres
On the Manner of Negotiating with Princes, 1716

✧✧✧

Yet what is all that fires a hero's scorn of Death?
The hope to live in hearts unborn.

- Thomas Campbell

✧✧✧

Victory has a hundred fathers, but defeat is an orphan.

- Thomas Campbell

✧✧✧

To live in hearts we leave behind is not to die.

- Thomas Campbell

✧✧✧

The side with the simplest uniforms wins.

- Major Mark Cancian

✧✧✧

Though the task of the Army is the defence of the country's borders, it does not mean that we should accomplish it by conducting a defensive war. We should aim at rapid defeat of the enemy by offensive action.

- General KP Candeth

✧✧✧

An officer is nothing without the soldiers.

- Field Marshal KM Cariappa (1899-1993)
First Indian Chief of the Indian Army

✧✧✧

In modern warfare, a large army is not sufficient; it needs industrial potential behind it. If the army is the first line of defence, the industry is the second.

- Field Marshal KM Cariappa

✧✧✧

Soldiers know the facility of wars to solve the internal problems. We ought to be ashamed that today they had more peace in war than peace in peace.

- Field Marshal KM Cariappa

✧✧✧

The Army is there to serve the Government of the day, and we should make sure that it does not get mixed up with party politics. A soldier is above politics and should not believe in caste or creed,

- Field Marshal KM Cariappa

✧✧✧

I am an Indian and to the last breath would remain an Indian. To me there are only two *Stans—Hindustan* (India) and *Foujistan* (the Army).

- Field Marshal KM Cariappa

✧✧✧

There are no good or bad units; there are only good or bad officers

- Field Marshal KM Cariappa

✧✧✧

The deepest truths lie buried in silence.

- Thomas Carlyle (1795-1881)

✧✧✧

Our main business is not to see dimly at a distance, but to do what clearly lies at hand.

- Thomas Carlyle

✧✧✧

A man lives by believing something; not by debating and arguing about many things

- Thomas Carlyle

✧✧✧

Conviction never so excellent, is worthless until it coverts itself into conduct.

- Thomas Carlyle

✧✧✧

The first duty of man is to conquer fear; he must get rid of it, he cannot act till then.

- Thomas Carlyle

✧✧✧

Genius is an infinite capacity for taking pains.

- Thomas Carlyle

✧✧✧

Imagination is a poor matter when it has to part company with understanding.

- Thomas Carlyle

✧✧✧

War is a quarrel between two thieves too cowardly to fight their own battle; therefore they take boys from one village and another village, stick them into uniforms, equip them with guns, and let them loose like wild beasts against one other.

- Thomas Carlyle

✧✧✧

The General Order is always:
......To manoeuvre in a body and on the attack.
......To maintain strict but not pettifogging discipline.
......To keep the troops constantly at the ready.
......To employ the utmost vigilance on sentry go.
......To use the bayonet on every possible occasion.
And to follow up the enemy remorselessly until he is utterly destroyed.

- Lazare Carnot (1753-1823)
French revolutionary, First Order of the Day,
2 February 1794, to army commanders.

✧✧✧

I hold before you my hand with each finger standing erect and alone, and as long as they are held thus, not one of the tasks that the hand may perform can be accomplished. I cannot lift. I cannot grasp. I cannot hold. I cannot even make an intelligible sign until my fingers organize and work together. In this we should also learn a lesson.

- George Washington Carver (1864-1943)

✧✧✧

The revolution... is a dictatorship of the exploited against the exploiters.

- Fidel Castro

✧✧✧

I warn you, I am just beginning! If there is in your hearts a vestige of love for your country, love for humanity, love for justice, listen carefully... I know that the regime will try to suppress the truth by all possible means; I know that there will be a conspiracy to bury me in oblivion. But my voice will not be stifled—it will rise from my breast even when I feel most alone, and my heart will give it all the fire that callous cowards deny it... Condemn me. It does not matter. History will absolve me.

- Fidel Castro
During his trial in 1953 when he was sentenced to fifteen years in prison.

✧✧✧

As I have said before, the ever more sophisticated weapons piling up in the arsenals of the wealthiest and the mightiest can kill the illiterate, the ill, the poor and the hungry, but they cannot kill ignorance, illness, poverty or hunger.

- Fidel Castro, 2002

✧✧✧

A revolution is not a bed of roses. A revolution is a struggle between the future and the past.

- Fidel Castro

✧✧✧

I began the revolution with 82 men. If I had to do it again, I will do it with 10 or 15 and absolute faith. It does not matter how small you are, if you have faith and plan of action.

- Fidel Castro

✧✧✧

Men do not shape destiny. Destiny produces the man for the hour.

- Fidel Castro

Marriage is an adventure, like going to war.

- Gilbert Keith Chesterton

✧✧✧

The Mauryan soldier does not himself the royal treasuries enrich nor does he the royal granaries fill. He does not himself carry out trade and commerce nor produce scholars, littérateurs, artistes, artisans, sculptors, architects, craftsmen, doctors and administrators. He does not himself build roads and ramparts nor dig wells and reservoirs. He does not himself write poetry or plays nor delve in metaphysics, arts and sciences. He does not do any of this directly.

The soldier only and merely ensures that the tax, tribute and revenue collectors travel far and wide and return safely; that the farmer tills, harvests, stores and markets his produce unafraid of pillage; that the trader, merchant and financier function and travel across the length and breadth of the realm unmolested; that the savant, sculptor, painter, maestro and mentor create works of art, literature, philosophy and astrology in quietitude; that the architect designs and builds his *vaastus* without tension; that the tutor and the priest teach and preach in peace; that the *rishis* meditate in wordless silence; that the *vaidya* (doctor) cures and invents medical formulations undisturbed; that the mason, the bricklayer, the artisan, the weaver, the tailor, the potter, the carpenter and the smith work unhindered; that the mother and the wife go about their chores and bring up children in harmony and tranquillity; that the aged fade away gracefully and with dignity; cattle graze freely without being lifted or stolen.

He is thus the very basis and silent, barely visible cornerstone of our fame, culture, physical well being and prosperity i.e. the entire nation building activity. He does not perform any of these chores himself directly: he enables the rest of us to perform these without let or hindrance.

Our military sinews, on the other hand, lend credibility to our adherence to good *dharma*, our goodwill, amiability and peaceful intentions towards all our neighbour nations as also those far away and beyond. These also serve as a powerful deterrent against military misadventure by any one of them.

If Pataliputra reposes each night in peaceful comfort, o king, it is so because it is secure in the belief that the distant borders of Magadha are inviolate and the interiors are safe and secure, thanks to the mighty Mauryan army standing vigil with naked swords and eyes peeled for action, day and night, in weather fair and foul, all eight *praharas* (dawn-to-dusk-to-dawn), quite unmindful of personal discomfort and hardship, loss of life and limb, separation from the family, all through the year, year after year.

While the citizenry of the state endeavours to make the state prosper and flourish, the Mauryan soldier guarantees it continues to exist.

To this man, *O Rajadhiraja*, you owe a debt for that very guarantee which is a vital *sine qua non* to our nationhood. Please, therefore, see to it, *suo motu,* that you are constantly alive and sensitive to the soldier's legitimate dues in every form and respect, be they his needs or his wants, including his place in the social order, and ensure that he receives them in time or ahead of time, in full measure, for he is not likely to ask for them himself.

This is so because before getting so completely wrapped up in his onerous, harsh and exalted charge, the soldier has assumed with good reason that the state, in return for his services, has freed him from all responsibility of his own present and future welfare as also that of his family back home in the hinterland. He is thus very clear in his mind when deployed at a distant border outpost or when campaigning in faraway lands that he need only look out in front for the enemy of the state and concentrate only on his military onus, completely free of all temporal worries. This assumption is a holy sacrament and an unwritten covenant that exists between him and the state. And rightly so!

If ever things come to a sordid pass when, on a given day, the Mauryan soldier has to look back over his shoulder (*simhawalokana*) prompted by even a single worry about his and his family's material, physical and social well being, it should cause you and your council the greatest concern and distress. I beseech you to take instant note and act with uncommon dispatch to address the soldier's anxiety. It is my bounden duty to assure you, my lord, that the day when the Mauryan soldier has to demand his dues or, worse, plead for them, will neither have arrived overnight nor in vain. It will also bode ill for Magadha.

For then, on that day, you, my lord, will have lost all moral sanction to be king! It will also be the beginning of the end of the Mauryan empire!!"

- *Chanakya (c. 350–283 BCE)*

❖❖❖

In the happiness of the people lies the ruler's happiness. Their welfare is his welfare. The ruler shall not consider what pleases and benefits him personally, but what is pleasing and beneficial to the people.

- Chanakya

✧✧✧

Exercise of power and achievement of results should be properly matched by the ruler in order to win over the people.

- Chanakya

✧✧✧

Miraculous results can be achieved by practicing the methods of subversion.

- Chanakya

✧✧✧

There are six major approaches to foreign policy. The first is a policy of maintaining peace with another state which is based on a treaty detailing the terms and conditions. The second is the policy of hostility which be followed if one is stronger than the enemy. The third approach is one of inaction: It is most suitable when states are of equal strength. The fourth is outright invasion but this policy is recommended for the very strong. For the very weak is prescribed the fifth approach, i.e., seeking shelter with another king and wait for better days, The sixth and the last approach recommends a policy of peace with one king/state while maintaining hostility towards another; such a dual policy is possible if help is available from another state to fight the enemy.

- Chanakya

✧✧✧

The international arena, (*mandala*) comprises 12 types of kings/states as follows:

1. The would-be conqueror, at the centre of the *mandala*.
2. The enemy whose territory borders on that of the would-be conqueror, i.e., the hostile neighbour.
3. The ally's whose territory lies immediately beyond that of the hostile neighbour.
4. The enemy's ally who is the neighbour of one's won ally.
5. The ally's ally who is territorially distant.
6. The ally of the enemy's ally who is also territorially distant.

Types 7 to 10 follow the same sequence but to the rear of the would-be conqueror.

11. A neutral king/state neighbouring both the would-be conqueror and his/its enemy but is stronger than both.
12. The king who is totally indifferent towards all other kings/states but is more powerful than the would-be conqueror, his enemy and the neutral king/state.

- Chanakya

✧✧✧

Many rulers have been destroyed by being under the control of the group of six enemies (lust, anger, greed, infatuation, arrogance, envy). Those with character should not follow their path, but preserve righteousness and wealth.

- Chanakya

✧✧✧

The people help the ruler who is just in his actions, when attacked while in a serious calamity.

- Chanakya

✧✧✧

The whole of this science is intended to create control over the senses. A ruler acting contrary to it and hence not having the senses under control immediately gets destroyed, even if he be the lord of the four ends of the earth.

- Chanakya

✧✧✧

A ruler with contiguous territory is a rival. The ruler next to the adjoining is to be deemed a friend.

- Chanakya

✧✧✧

A mad elephant, mounted by an intoxicated mahout tramples on everything that it comes across. Likewise is a ruler devoid of the light of learning, advised by an unwise minister.

- Chanakya

✧✧✧

A ruler, disciplined by learning, will be interested in disciplining his subjects. He will enjoy the earth unopposed, devoted to the welfare of all beings

- Chanakya

✧✧✧

One who does not keep one's word and one whose behaviour is contrary to that of the people becomes untrustworthy to one's own people and to others. Hence {a ruler} should adopt the same mode of life, same dress, same language, and same customs as those of the people.

- Chanakya

✧✧✧

A ruler with loyal people accomplishes his task even with a little help because of their cooperation.

- Chanakya

✧✧✧

A ruler who administers justice on the basis of four principles: righteousness, evidence, history of the case and the prevalent law shall conquer the four corners of the earth.

- Chanakya

✧✧✧

A ruler with character can render even unendowed people happy. A characterless ruler destroys loyal and prosperous people.

- Chanakya

✧✧✧

Power is the cause for the forging of treaties. Unheated metal does not join with metal.

- Chanakya

✧✧✧

The force of (an army) which returns to fight, without any desire to live, is irresistible. Hence a broken army should, not be harassed.

- Chanakya

✧✧✧

In daytime the crow kills the owl. At night the owl kills the crow. (The time of fight is important.)

- Chanakya

✧✧✧

The policy of *vigraha* (state of hostility)is to be followed if (the ruler) were to see that, "my country, consisting of martial people or fighting bands, or secure in the protection of a single entrance through a mountain fort, a forest fort or a river fort, will be able to repulse the enemy's attack; or, taking shelter in an impregnable fort on the border of my territory, I shall be able to ruin the enemy's undertakings; or, the enemy, with his energy sapped by the troubles caused by a calamity, has reached a time when his undertakings face ruin; or, when he is fighting elsewhere, I shall be able to carry off his country," he should secure advancement by resorting to war.

- Chanakya

✧✧✧

Before you start some work, always ask yourself three questions—Why am I doing it? What the results might be? and Will I be successful? Only when you think deeply and find satisfactory answers to these questions, go ahead.

- Chanakya

✧✧✧

The serpent, the king, the tiger, the stinging wasp, the small child, the dog owned by other people, and the fool: these seven ought not to be awakened from sleep.

- Chanakya

✧✧✧

An army not under control can be brought under control by conciliation and other means. Wealth is the key to raising successful armies and having a peaceful and just kingdom, and Kautilya's political science brings wealth.

- Chanakya

✧✧✧

The arrow shot by the archer may or may not kill a single person. But stratagems devised by wise men can kill even babes in the womb.

- Chanakya

✧✧✧

A dog on land drags a crocodile. A crocodile in water drags a dog. (The place of fight is important.)

- Chanakya

✧✧✧

The safety, honour and welfare of your country come first, always and every time; The honour, welfare and comfort of the men you command come next; Your own ease, comfort and safety come last, always and every time.

- Field Marshal Philip Chetwode
Address at the inauguration of the Indian Military Academy, Dehradun, 10 December 1932

✧✧✧

Attacking does not merely consist in assaulting walled cities or striking at an army in battle array; it must include the art of assailing the enemy's mental equilibrium.

- Li Ching (571-649 C.E.)

✧✧✧

Terrorism has become the systematic weapon of a war that knows no borders or seldom has a face.

- Jacques Chirac, 24 September 1986

✧✧✧

Wanton killing of innocent civilians is terrorism, not a war against terrorism.

- Noam Chomsky, 9-11

✧✧✧

A nation can survive its fools and even the ambitious. But it cannot survive treason from within. An enemy at the gates is less formidable, for he is known and he carries his banners openly against the city. But the traitor moves among those within the gates freely, his sly whispers rustling through all alleys, heard in the very halls of government itself. For the traitor appears no traitor; he speaks in the accents familiar to his victim, and he wears their face and their garments and he appeals to the baseness that lies deep in the hearts of all men. He rots the soul of a nation; he works secretly and unknown in the night to undermine the pillars of a city; he infects the body politic so that it can no longer resist. A murderer is less to be feared. The traitor is the plague.

- Marcus Tullius Cicero, Roman Orator—106-43 B.C.

✧✧✧

I would remind you that extremism in the defence of liberty is no vice.

- Marcus Tullius Cicero

✧✧✧

An unjust peace is better than a just war.

- *Marcus Tullius Cicero*

✧✧✧

Any man can make mistakes, but only an idiot persists in his error.

- *Marcus Tullius Cicero*

✧✧✧

Before beginning, plan carefully.

- *Marcus Tullius Cicero*

✧✧✧

If you have no confidence in self, you are twice defeated in the race of life. With confidence, you have won even before you have started.

- *Marcus Tullius Cicero*

✧✧✧

In time of war the laws are silent.

- *Marcus Tullius Cicero*

✧✧✧

History is indeed the witness of the times, the light of truth.

- *Marcus Tullius Cicero*

✧✧✧

The only excuse, therefore, for going to war is that we may live in peace unharmed.

- *Marcus Tullius Cicero*

✧✧✧

A letter does not blush.

- *Marcus Tullius Cicero*

✧✧✧

A man of courage is also full of faith.

- *Marcus Tullius Cicero*

✧✧✧

Any man is liable to err, only a fool persists in error.

- *Marcus Tullius Cicero*

✧✧✧

Glory follows virtue as if it were its shadow.

- *Marcus Tullius Cicero*

✧✧✧

Rashness belongs to youth; prudence to old age.

- *Marcus Tullius Cicero*

✧✧✧

Probabilities direct the conduct of the wise man.

- *Marcus Tullius Cicero*

✧✧✧

The authority of those who teach is often an obstacle to those who want to learn.

- *Marcus Tullius Cicero*

✧✧✧

The enemy is within the gates; it is with our own luxury, our own folly, our own criminality that we have to contend.

- Marcus Tullius Cicero

✧✧✧

An army is of little use in the field unless there are wise counsels at home.

- Cicero

✧✧✧

The US Army will never control the ground under the sky, if the US Air Force does not control the sky over the ground.

- Col Gene Cirillo, USAF (Ret)

✧✧✧

Without supplies neither a general nor a soldier is good for anything.

- Clearchus, 401 BC. Spartan officer,
celebrated as the leader of the Ten Thousand.

✧✧✧

The nation which forgets its defenders will be itself forgotten.

- Calvin Coolidge (1872-1933)

✧✧✧

Why don't we just buy one airplane and let the pilots take turns flying it.

- Calvin Coolidge
on a War Department request to buy more aircraft

✧✧✧

Now I recall the Recon Marines ragged, filthy cammie shirted young men in green paint who move silent like the fog with deadly purpose in their eyes. Swift, Silent, Deadly. I smile.

- Sergeant-Major Correll, US Marine Corps

✧✧✧

Gentlemen, we are being killed on the beaches. Let's go inland and be killed.

- Brigadier-General Norman Cota
Omaha Beach, 6 June 1944

✧✧✧

We are outnumbered, there is only one thing to do. We must attack!

- Admiral Andrew Cunningham, 11 November 1940

✧✧✧

There is the guilt all soldiers feel for having broken the taboo against killing, a guilt as old as war itself. Add to this the soldier's sense of shame for having fought in actions that resulted, indirectly or directly, in the deaths of civilians. Then pile on top of that an attitude of social opprobrium, an attitude that made the fighting man feel personally morally responsible for the war, and you get your proverbial walking time bomb.

- Philip Caputo, 1982

✧✧✧

We must conquer war or war will conquer us.

- Ely Culbertson

✧✧✧

Patriotism is not enough. I must have no hatred or bitterness towards anyone.

- Edith Cavell (1865–1915) British nurse and humanitarian. She saved the lives of casualties from all sides without distinction and helped some 200 Allied soldiers escape from German-occupied Belgium during World War I, for which she was arrested, court-martialled and found guilty of treason. She was sentenced to death and shot by firing squad.

✧✧✧

He who loses wealth loses much; he who loses a friend loses more; but he who loses courage loses all.

- Miguel de Cervantes (1547–1616), Spanish novelist, poet, and playwright.

✧✧✧

It is one thing to praise discipline, and another to submit to it.

- Miguel de Cervantes

✧✧✧

A proverb is a short sentence based on long experience.

- Miguel de Cervantes

✧✧✧

Never stand begging for that which you have the power to earn.

- Miguel de Cervantes

✧✧✧

However much we may sympathize with a small nation confronted by a big and powerful neighbour, we cannot in all circumstances undertake to involve the whole British Empire in a war simply on her account. If we have to fight it must be on larger issues than that...

- Neville Chamberlain (1938) on the conflict between Germany and Czechoslovakia

✧✧✧

Whatever the lengths to which others may go, His Majesty's Government will never resort to the deliberate attack on women and children and other civilians for purposes of mere terrorism.

- British Prime Minister Neville Chamberlain, House of Commons, 14 September 1939.

✧✧✧

Against this pale, duck-egg blue and greyish mauve were silhouetted a number of small black shapes: all of them bombers, and all of them moving the same way. One hundred and thirty-four miles ahead, and directly in their path, stretched a crimson-red glow; cologne was on fire. Already, only twenty-three minutes after the attack had started, Cologne was ablaze from end to end, and the main force of the attack was still to come.

- Group Captain Leonard Cheshire, VC, DSO

✧✧✧

Though all under heaven be at peace, if the art of war be forgotten there is peril.

- Chinese proverb

◇◇◇

True gold does not fear the test of fire.

- Chinese proverb

◇◇◇

In war, resolution;
In defeat, defiance;
In victory, magnanimity.

- Winston Churchill

◇◇◇

The peoples, transported by their sufferings and by the mass teachings with which they had been inspired, stood around in scores of millions to demand that retribution should be exacted to the full. Woe betide the leaders now perched on their dizzy pinnacles of triumph if they cast away at the conference table what the soldiers had won on the hundred blood-soaked battlefields ... The Multitudes remained plunged in ignorance of the simplest facts, and their leaders, seeking their votes, did not dare to undeceive the.

- Winston Churchill
writing on the Paris Peace conference, 1919

◇◇◇

But nothing daunted the valiant heart of man. Son of the Stone Age, vanquisher of nature with all her trials and monsters, he met the awful and self-inflicted agony with new reserves of fortitude. Freed in the main by his intelligence from mediaeval fears, he marched to death with sombre dignity. His nervous system was found in the Twentieth Century capable of enduring physical and moral stresses before which the simpler natures of primeval times would have collapsed. Again and again to the hideous bombardment, again and again from the hospital to the front, again and again to the hungry submarines, he strode unflinching. And withal, as an individual, preserved through these torments the glories of a reasonable and compassionate mind.

- Winston Churchill, The World Crisis 1911-1914

◇◇◇

Success is not final, failure is not fatal: it is the courage to continue that counts.

- Sir Winston Churchill

◇◇◇

Regiments are not like houses. they cannot be pulled down and altered structurally to suit the convenience of the occupier or the caprice of the owner. They are more like plants; they grow slowly if they are to grow strong...and if they are blighted or transplanted they are apt to wither.

- Winston Churchill, 1904

◇◇◇

Please see 'The Times' of February 4. Is it really true that a seven-mile cross-country run is enforced upon all in this division from generals to privates? Does the Army Council think this is a good idea? It looks to me rather excessive. A colonel or a general ought not to exhaust himself in trying to compete with young boys running across the country seven miles at a time. The duty of officers is no doubt to keep themselves fit, but still more to think for their men and to take decisions affecting their safety and comfort. Who is the general of this division, and does he run the seven miles himself? If so, he may be more useful for football than war. Could Napoleon have run seven miles across the country at Austerlitz? Perhaps it was the other fellow he made run. In my experience, based on many years' observation, officers with high athletic qualifications are not usually successful in the higher ranks.

-Winston Churchill
(Minute to the Secretary of State for War, 4 February 1941)

✧✧✧

It is always wise to look ahead, but difficult to look further than you can see.

- Winston Churchill

✧✧✧

Nearly all battles which are regarded as masterpieces of the military art, from which have been derived the foundation of states and the fame of commanders, have been battles of manoeuvre in which the enemy has found himself defeated by some novel expedient or device, some queer, swift, unexpected thrust or stratagem. In many battles the losses of the victors have been small. There is required for the composition of the great commander not only massive common sense and reasoning power, not only imagination, but also an element of the legerdemain, an original and sinister touch, which leaves the enemy puzzled as well as beaten. It is because military leaders are credited with gifts of this order which enable them to ensure victory and save slaughter that their profession is held in such high honour.....

- Winston Churchill, The World Crisis: 1914-1918

✧✧✧

Courage is what it takes to stand up and speak; courage is also what it takes to sit down and listen.

- Winston Churchill

✧✧✧

Attitude is a little thing that makes a big difference.

- Winston Churchill

✧✧✧

Renown awaits the commander who first restores artillery to its prime importance on the battlefield.

- Winston Churchill

✧✧✧

In one phase men seem to have been right, in another they seem to have been wrong. Then again, a few years later, when the perspective of time has lengthened, all stands in a different setting. There is a new proportion. There is another scale of values. History with its flickering lamp stumbles along the trail of the past, trying to reconstruct its scenes, to revive its echoes, and kindle with pale gleams the passion of former days.

- Winston Churchill,
Tribute to Neville Chamberlain: A Speech to The House of Commons, 12 November 1940

✧✧✧

An appeaser is one who feeds a crocodile, hoping it will eat him last.

- Winston Churchill

✧✧✧

I am easily satisfied with the very best.

- Winston Churchill

✧✧✧

Never, never, never believe any war will be smooth and easy, or that anyone who embarks on the strange voyage can measure the tides and hurricanes he will encounter. The statesman who yields to war fever must realize that once the signal is given, he is no longer the master of policy but the slave of unforeseeable and uncontrollable events.

- Sir Winston Churchill (1874–1965)

✧✧✧

For good or for ill, air mastery is today the supreme expression of military power and fleets and armies, however vital and important, must accept a subordinate rank.

- Prime Minister Winston Churchill

✧✧✧

What General Weygand called the Battle of France is over. I expect the Battle of Britain is about to begin . . Let us therefore brace ourselves to our duties, and so bear ourselves that, if the British Empire and its Commonwealth last for a thousand years, men will still say, "This was their finest hour."

- Prime Minister Winston Churchill, 18 June 1940

✧✧✧

Never in the field of human conflict was so much owed by so many to so few.

- Prime Minister Winston Churchill, House of Commons, 20 August 1940. The Royal Air Force has been known as 'the few' ever since. M. Hastings (2009) Winston's War states that Churchill came up with the phrase a few days earlier on 16 August, after visiting Fighter Command's 11 Group operations room. His chief of staff 'Pug' Ismay made some remark in the car riding back to Chequers, and Churchill said, "Don't speak to me. I have never been so moved." After a few minutes he spoke the classic line.

✧✧✧

The empires of the future are empires of the mind.

- Winston Churchill

✧✧✧

. . . when I look round to see how we can win the war I see that there is only one sure path . . . and that is absolutely devastating, exterminating attack by very heavy bombers from this country upon the Nazi homeland. We must be able to overwhelm them by this means, without which I do not see a way through.

- British Prime Minister Winston Churchill in a letter to Minister of Aircraft Production Lord Beaverbrook, July 1940.

✧✧✧

Success is not final, failure is not fatal: it is the courage to continue that counts.

- Winston Churchill

✧✧✧

The Navy can lose us the war, but only the Air Force can win it. Therefore our supreme effort must be to gain overwhelming mastery in the Air. The Fighters are our salvation . . . but the Bombers alone provide the means of victory. . . . In no other way at present visible can we hope to overcome the immense military power of Germany.

- Prime Minister Winston Churchill memorandum for the Cabinet, 3 September 1940.

✧✧✧

If you are going through hell, keep going.

- Winston Churchill

✧✧✧

Courage is the first of human qualities because it is the quality which guarantees all others.

- Winston Churchill

✧✧✧

Never turn your back on a threatened danger and try to run away from it. If you do that, you will double the danger. But if you meet it promptly and without flinching, you will reduce the danger by half. Never run away from anything. Never!

- Winston Churchill

✧✧✧

The military code ...... has come down to us from even before the age of knighthood and chivalry. It embraces the highest moral laws, and will stand the test of any ethic or philosophies ever promulgated for the upliftment of mankind. Its requirements are for the things that are right, and its restraints are from things that are wrong. Its observance will uplift everyone who comes under its influence. The soldier, above all other men, is required to perform the highest act of religious teaching—sacrifice.

- Winston Churchill

✧✧✧

In war, truth must have an escort of lies.

- Winston Churchill

✧✧✧

However horrible the incidents of war may be, the soldier who is called upon to offer and to give his life for his country is the noblest development of mankind.

- Winston Churchill

✧✧✧

The belief that security can be obtained by throwing a small state to the wolves is a fatal delusion.

- Winston Churchill

✧✧✧

Air power can either paralyze the enemy's military action or compel him to devote to the defence of his bases and communications a share of his straitened resources far greater that what we need in the attack.

- Prime Minister Winston Churchill

✧✧✧

But after all, the great defence against the air menace is to attack the enemy's aircraft as near as possible to their point of departure.

- Winston Churchill, 1914

✧✧✧

Air power may either end war or end civilization.

- Winston Churchill

✧✧✧

For good or for ill, air mastery is today the supreme expression of military power and fleets and armies, however vital and important, must accept a subordinate rank.

- Winston Churchill, 1949

✧✧✧

Not to have an adequate air force in the present state of the world is to compromise the foundations of national freedom and independence.

- Winston Churchill, House of Commons, 14 March 1933

✧✧✧

It is dangerous to meddle with Admirals when they say they can't do things. They have always got the weather or fuel or something to argue about.

- Winston Churchill

✧✧✧

However absorbed a commander may be in the elaboration of his own thoughts, it is necessary sometimes to take the enemy into consideration.

- Winston Churchill

✧✧✧

Victory was to be bought so dear as to be almost indistinguishable from defeat.

- Winston Churchill

✧✧✧

I would say to the House, as I said to those who have joined this government: "I have nothing to offer but blood, toil, tears and sweat." We have before us an ordeal of the most grievous kind. We have before us many, many long months of struggle and of suffering. You ask, what is our policy? I will say; "It is to wage war, by sea, land and air, with all our might and with all the strength that God can give us: to wage war against a monstrous tyranny, never surpassed in the dark lamentable catalogue of human crime. That is our policy." You ask, what is our aim? I can answer with one word: Victory—victory at all costs, victory in spite of all terror, victory however long and hard the road may be; for without victory there is no survival."

- Winston Churchill
(First Address as Prime Minister, 13 May 1940)

✧✧✧

We shall not flag nor fail. We shall go on to the end. We shall fight in France and on the seas and oceans; we shall fight with growing confidence and growing strength in the air. We shall defend our island whatever the cost may be; we shall fight on beaches, landing grounds, in fields, in streets and on the hills. We shall never surrender and even if, which I do not for the moment believe, this island or a large part of it were subjugated and starving, then our empire beyond the seas, armed and guarded by the British Fleet, will carry on the struggle until in God's good time the New World with all its power and might, sets forth to the liberation and rescue of the Old."

- Winston Churchill, 4 June 1940

✧✧✧

Here is the answer which I will give to President Roosevelt: We shall not fail or falter; we shall not weaken or tire... Neither the sudden shock of battle nor the long-drawn trials of vigilance and exertion will wear us down. Give us the tools and we will finish the job."

- Winston Churchill, Radio broadcast, 9 February 1941

✧✧✧

In wartime, truth is so precious that she should always be attended by a bodyguard of lies.

- Winston Churchill

✧✧✧

If you will not fight for the right when you can easily win without bloodshed, if you will not fight when your victory will be sure and not too costly, you may come to the moment when you will have to fight with all the odds against you and only a small chance of survival. There may even be a worse case: you may have to fight when there is no hope of victory, because it is better to perish than to live as slaves.

- Winston Churchill

✧✧✧

Nations which go down fighting rise again, those who surrender tamely are finished.

- Winston Churchill

✧✧✧

It is probable that future war will be conducted by a special class, the Air Force, as it was by the armoured knights of the Middle Ages.

- Winston Churchill

✧✧✧

Never in the field of human conflict was so much owed by so many to so few.

- Winston Churchill (about the Royal Air Force)

✧✧✧

No matter how enmeshed a commander becomes in the elaboration of his own thoughts, it is sometimes necessary to take the enemy into account.

- Winston Churchill

✧✧✧

The Americans will always do the right thing... after they've exhausted all the alternatives.

- Winston Churchill

✧✧✧

We have a very daring and skilful opponent against us, and, may I say across the havoc of war, a great general.

- Churchill about Rommel

✧✧✧

It seems to me that the moment has come when the question of bombing of German cities simply for the sake of increasing the terror, though under other pretexts, should be reviewed. Otherwise we shall come into control of an utterly ruined land . . . The destruction of Dresden remains a serious query against the conduct of Allied bombing . . . I feel the need for more precise concentration upon military objectives, such as oil and communications behind the immediate battle-zone, rather than on mere acts of terror and wanton destruction, however impressive.

- Winston Churchill, British Prime Minister, memo to Charles Portal, Chief of the Air Staff and the Chiefs of Staff Committee, 28 March 1945.

Under pressure from Air Chief Marshal Arthur Harris, Portal and others, Churchill withdrew his memo and issued a new one on 1 April 1945 omitting the words "acts of terror."

✧✧✧

Not to have an adequate air force in the present state of the world is to compromise the foundations of national freedom and independence.

- Winston Churchill, House of Commons, 14 March 1933.

✧✧✧

We cannot guarantee victory, but only deserve it.

- Winston Churchill

✧✧✧

Courage is grace under pressure.

- Sir Winston Churchill

✧✧✧

Air power may either end war or end civilization.

- Winston Churchill, House of Commons, 14 March 1933.

✧✧✧

They had bombed London, whether on purpose or not, and the British people and London especially should know that we could hit back. It would be good for the morale of us all.

- Winston Churchill, British Prime Minister, ordering the RAF to start bombing German cities, cited in M. Hastings, Winston's War, 25 August 1940.

✧✧✧

From now on we shall bomb Germany on an ever-increasing scale, month by month, year by year, until the Nazi regime has either been exterminated by us or—better still—torn to pieces by the German people themselves.

- Prime Minister Winston Churchill, 14 July 1941.

✧✧✧

"lions led by donkeys"

- Alan Clark

a phrase from the book, 'The Donkeys' popularly used to describe the British infantry of the First World War and to condemn the generals who commanded them. It is attributed it to a conversation between German generals Erich Ludendorff and Max Hoffmann:

Ludendorff: The English soldiers fight like lions.

Hoffmann: True. But don't we know that they are lions led by donkeys.

✧✧✧

You must be able to underwrite the honest mistakes of your subordinates if you wish to develop their initiative and experience.

- General Bruce C. Clarke

✧✧✧

Rank is only given you in the Army to enable you to better serve those below you and those above you. Rank is not given for you to exercise your idiosyncrasies.

- General Bruce C. Clarke

✧✧✧

When things go wrong in your command, start searching for the reason in increasingly larger concentric circles around your own desk.

- General Bruce C. Clarke

✧✧✧

Modern leadership demands officers who can accept challenge with initiative, originality, fidelity, understanding, and, above all, the willingness to fully assume the responsibilities of command.

- General Bruce C. Clarke

✧✧✧

One of the principles that we operate on in this country is that leaders are held accountable. The simple truth is that we went into Iraq on the basis of some intuition, some fear, and some exaggerated rhetoric and some very, very scanty evidence."

- General Wesley Clark

✧✧✧

It's essential to have boots and eyes on the ground.

- General Wesley Clark

✧✧✧

Serious research and development efforts are required to produce technologies, strategies, organizations, and trained personnel who can go into failed states, work with our allies and friends, and promote the political and economic reforms that will meet popular needs and reduce the sources of terrorism and conflict.

- Wesley Clark, Winning Modern Wars

✧✧✧

Defeating terrorism is more difficult and far-reaching than we have assumed.... We may be advancing the ball down the field at will, running over our opponent's defences, but winning the game is another matter altogether.

- Wesley Clark, Winning Modern Wars

✧✧✧

The first and most important rule to observe is to use our entire forces with the utmost energy. The second rule is to concentrate our power as much as possible against that section where the chief blows are to be delivered and to incur disadvantages elsewhere, so that our chances of success may increase at the decisive point. The third rule is never to waste time, unless important advantages are to be gained from hesitation, it is necessary to set to work at once. By this speed a hundred enemy measures are nipped in the bud, and public opinion is won most rapidly. Finally, the fourth rule is to follow up our successes with the utmost energy. Only pursuit of the beaten enemy gives the fruits of victory.

- Karl von Clausewitz

✧✧✧

The soldier's trade, if it is to mean anything at all, has to be anchored to an unshakable code of honour. Otherwise, those of us who follow the drums become nothing more than a bunch of hired assassins walking around in gaudy clothes . . . a disgrace to God and mankind.

- Karl von Clausewitz, 1832

✧✧✧

There is only one decisive victory: the last.

- Karl von Clausewitz

✧✧✧

The end for which a soldier is recruited, clothed, armed, and trained, the whole object of his sleeping, eating, drinking, and marching is simply that he should fight at the right place and the right time.

- Karl von Clausewitz

✧✧✧

We see, therefore, that War is not merely a political act, but also a real political instrument, a continuation of political commerce, a carrying out of the same by other means. All beyond this which is strictly peculiar to War relates merely to the peculiar nature of the means which it uses. . . But however powerfully this may react on political views in particular cases, still it must always be regarded as only a modification of them; for the political view is the object, War is the means, and the means must always include the object in our conception."

- Karl von Clausewitz, "Vom Kriege" (On War)

✧✧✧

In war, more than anywhere else in the world, things happen differently to what we had expected .... the commander is a victim of a hundred thousand impressions, of which the most have an intimidating, the fewest an encouraging, tendency .... therefore, only an immense force of will, which manifests itself in perseverance, can conduct us to the aim.

- Karl von Clausewitz

✧✧✧

If the leader is filled with high ambition and if he pursues his aims with audacity and strength of will, he will reach them in spite of all obstacles.

- Karl von Clausewitz

✧✧✧

All military action is permeated by intelligent forces and their effects.

- Karl von Clausewitz

✧✧✧

A good deal of information gathered during war is contradictory, a still greater part of it erroneous and the bulk of military information s of questionable reliability. Plans which are built on such grounds may fail.

- Karl von Clausewitz

✧✧✧

The moral elements are amongst the most important in war. They constitute the spirit that permeates war as a whole and at an early stage they establish a close affinity with the will that moves and leads the whole mass of force, practically merging with it, since the will is itself a moral quality .... History provides the strongest proof of the importance of moral factors and their often incredible effects .....

- Karl von Clausewitz

✧✧✧

To bring a war, or one of its campaigns, to a successful conclusion requires a thorough grasp of national policy. On that level, strategy and policy coalesce: the commander-in-chief is simultaneously a statesman.

- Karl von Clausewitz

✧✧✧

If every soldier needed some degree of military genius our armies would be very weak, for the term refers to a special cast of mental or moral powers which can rarely occur in an army when a society has to employ its abilities in many different areas. The smaller the range of activities of a nation and the more the military factor dominates, the greater will be the incidence of military genius. This however is true only of its distribution, not of its quality. The latter depends on the general intellectual development of a given society. In any primitive, warlike race, the warrior spirit is far more common than among civilized peoples. It is possessed by almost every warrior: but in civilized societies only necessity will stimulate it in the people as a whole, since they lack the natural disposition for it.

- Karl von Clausewitz, On War

✧✧✧

We are therefore certain that no rules of any kind exist for manoeuvre, and no method or general principle can determine the value of the action, rather, superior application, precision, order, discipline, and fear will find the means to achieve palpable advantage in the most singular and minute circumstances. It is on these qualities that victory in this type of contest largely depends.

- Karl von Clausewitz, On War

✧✧✧

Four elements make up the climate of war: danger, exertion, uncertainty and chance. If we consider them together, it becomes evident how fortitude of mind and character are needed to made progress in these impeding elements with safety and success. According to circumstance, reporters and historians of war use such terms as energy, firmness, staunchness, emotional balance, and strength of character. These products of a heroic nature could almost be treated as one and the same force—strength of will—which adjusts itself to circumstances: but though closely linked, they are not identical. A closer study of the interplay of psychological forces at work here may be worthwhile.

- Karl von Clausewitz, On War

✧✧✧

The end for which a soldier is recruited, clothed, armed, and trained, the whole object of his sleeping, eating, drinking, and marching is simply that he should fight at the right place and the right time.

- Karl von Clausewitz

✧✧✧

Victory normally results from the superiority of one side; from a greater aggregate of physical and psychological strength. This superiority is certainly augmented by victory, otherwise it would not be so coveted or command so high a price. That is an automatic consequence of victory itself. Its effects exert a similar influence, but only up to a point. That point may be reached quickly-at times so quickly that the total consequences of a victorious battle may be limited to an increase in psychological superiority alone.

- Karl von Clausewitz, 1780–1831

✧✧✧

One must keep the dominant characteristics of both belligerents in mind. Out of these characteristics a certain centre of gravity develops, the hub of all power and movement, on which everything depends. That is the point against which all our energies must be directed.

- Karl von Clausewitz

✧✧✧

War is much too serious a matter to be entrusted to the military.

- George Clemenceau

✧✧✧

If only we learn to deal with ourselves with our heads and others with our hearts, the world will be a great place to live in.

- Bill Clinton

✧✧✧

Small war almost always involves political interference in the affairs of the country in which it is waged; it is in the very nature of such wars that the military problems are difficult to distinguish from the political ones. The skills of manipulation which successful coalition warfare in such circumstances requires are not only scarce, but in some measure anathema to the American military. The desire of the American military to handle only pure "military" problems is ... understandable in light of its Vietnam experience, but unrealistic nonetheless.

- Eliot Cohen

✧✧✧

Physical courage, which despises all danger, will make a man brave in one way; and moral courage, which despises all opinion, will make a man brave in another. The former would seem most necessary for the camp, the latter for the council; but to constitute a great man both are necessary.

- Colton

✧✧✧

If I am walking with two other men, each of them will serve as my teacher. I will pick out the good points of the one and imitate them, and the bad points of the other and correct them in myself.

- Confucius

✧✧✧

He who learns but does not think, is lost! He who thinks but does not learn is in great danger.

- Confucius

✧✧✧

If we don't know life, how can we know death?

- Confucius

✧✧✧

The superior man, when resting in safety, does not forget that danger may come. When in a state of security, he does not forget the possibility of ruin. When all is orderly, he does not forget that disorder may come. Thus, his person is not endangered; his states and all his clans are preserved.

- Confucius

✧✧✧

I hear, I know. I see, I remember. I do, I understand.

- Confucius

✧✧✧

It does not matter how slowly you go as long as you do not stop.

- Confucius

✧✧✧

Our greatest glory is not in never falling, but in rising every time we fall.

- Confucius

✧✧✧

Real knowledge is to know the extent of one's ignorance.

- Confucius

✧✧✧

Study the past, if you would divine the future.

- Confucius

✧✧✧

Success depends upon previous preparation, and without such preparation there is sure to be failure.

- Confucius

✧✧✧

The cautious seldom err.

- Confucius

✧✧✧

The expectations of life depend upon diligence; the mechanic that would perfect his work must first sharpen his tools.

- Confucius

✧✧✧

The strength of a nation derives from the integrity of the home.

- Confucius

✧✧✧

The superior man acts before he speaks, and afterwards speaks according to his action.

- Confucius

✧✧✧

The superior man is distressed by the limitations of his ability; he is not distressed by the fact that men do not recognize the ability that he has.

- Confucius

✧✧✧

The superior man is modest in his speech, but exceeds in his actions.

- Confucius

✧✧✧

The superior man makes the difficulty to be overcome his first interest; success only comes later.

- Confucius

✧✧✧

The superior man thinks always of virtue; the common man thinks of comfort.

- Confucius

✧✧✧

The superior man understands what is right; the inferior man understands what will sell.

- Confucius

✧✧✧

The will to win, the desire to succeed, the urge to reach your full potential... these are the keys that will unlock the door to personal excellence.

- Confucius

✧✧✧

There are three methods to gaining wisdom. The first is reflection, which is the highest. The second is limitation, which is the easiest. The third is experience, which is the bitterest.

- Confucius

✧✧✧

To be wronged is nothing unless you continue to remember it.

- Confucius

✧✧✧

Leadership is a matter of correctness. If you lead by going down a correct path yourself, who will dare to take an incorrect one?

- Confucius

✧✧✧

When it is obvious that the goals cannot be reached, don't adjust the goals, adjust the action steps.

- Confucius

✧✧✧

To see what is right and not do it is want of courage.

- Confucius

✧✧✧

Those who wish to bring order to the State should first regulate their families. Those who wish to regulate their families should first cultivate their personal lives. Those who wish to cultivate their personal lives should first rectify their minds. Those who wish to rectify their minds should first make their wills sincere. Those who wish to make their wills sincere should first extend their knowledge.

- Confucius

✧✧✧

National military forces will become more like constabulary forces, albeit at the high end of technology and power. But like any constabulary, their operations will be increasingly constrained by law (or acceptable practice). Their targets will be less predictable and may include nation states, sub-national political organisations or mere criminals.

- M. O'Connor

✧✧✧

It is not possible to appreciate courage until the danger that has brought it into being has been experienced.

- Jim Corbett

✧✧✧

Discipline being the principal strength of all armies, it is essential that all superiors receive from their subordinates absolute obedience and submission on all occasions. Orders must be executed instantly without hesitation or complaint. The authorities who give them are responsible for them and an inferior is only permitted to make an objection after he has obeyed.

- A.R. Cooper, March or Bust;
Adventures in the Foreign Legion, 1972

✧✧✧

When a nation shows a civilized horror of war, it receives directly the punishment of its mistake. God changes its sex, despoils it of its common mark of virility, changes it into a feminine nation and sends conquerors to ravish it of its honour.

- Donoso Cortes

✧✧✧

To set and work toward any goal is an act of courage.

- Stephen R. Covey

✧✧✧

You cannot choose your battlefield,
God does that for you;
But you can plant a standard
Where a standard never flew.

- Nathalia Crane

✧✧✧

A man feared that he might find an assassin;
Another that he might find a victim.
One was wiser than the other.

- Stephen Crane

✧✧✧

He wishes that he, too, had a wound, a red badge of courage.

- Stephen Crane

✧✧✧

The Six Rules of Soldiering:

- Develop individual initiative and responsibility
- Never follow orders blindly
- Develop proper discipline
- Develop primary groups
- Develop an unremitting attack philosophy
- "The Golden Rule"—It is better to do something wrong but decisive, than to wait for orders which may never arrive.

- Martin van Creveld
(in a German Army Manual—Command of Troops, 1936)

✧✧✧

The Israeli army, which used to be that of a small state fighting much more powerful and larger opponents, has shifted to fighting opponents which are much smaller than itself.

- Martin van Creveld

✧✧✧

If you fight opponents who are weaker than yourself for any length, then you will find yourself disintegrating. It happened to the French in Algeria. It happened to the Americans in Vietnam. It happened to the Soviets in Afghanistan, and now it's happening to us

- Martin van Creveld

✧✧✧

Assuming China does not become destabilized and continues to grow, it will no doubt develop a military program in proportion to its resources.

- Martin van Creveld

✧✧✧

Except when war is waged in a desert, non-combatants, also known as civilians or "the people," constitute the great majority of those affected.

- Martin van Creveld

✧✧✧

It is simply not true that war is solely a means to an end, nor do people necessarily fight in order to obtain this objective or that. In fact, the opposite is true: people very often take up one objective or another precisely in order that they may fight.

- Martin van Creveld

✧✧✧

Sun Tzu does not need my praise. His work has lived for over two thousand years, and will surely live for another two thousand without any help from me.

- Martin van Creveld

✧✧✧

The enemy resembles us. Therefore, he needs to be approached not as an assembly of 'targets' to be destroyed one by one; but as a living, intelligent entity capable of acting and reacting.

- Martin van Creveld

✧✧✧

The problem is that you cannot prove yourself against someone who is much weaker than yourself. They are in a lose/lose situation. If you are strong and fighting the weak, then if you kill your opponent then you are a scoundrel... if you let him kill you, then you are an idiot. So here is a dilemma which others have suffered before us, and for which as far as I can see there is simply no escape."

- Martin Van Creveld
(The Changing Face of War: Lessons of Combat)

✧✧✧

Put your trust in God, but keep your powder dry.

- Oliver Cromwell, 1599-1658, Lord Protector of England

✧✧✧

I am the bomber 17—
Proud machine—sleek and powerful,
Made by man to kill his foe,
Made of steel and wood and metal,
Built to fight and drop destruction . . .

- Robert Cromwell, 'Skyward: A Ballad of the Bomber.'

✧✧✧

The greatest danger to the Staff College is a swollen head.

- F.P. Crozier (Impressions and Recollections)

✧✧✧

No speech of admonition can be so fine that it will at once make those who hear it good men if they are not good already; it would surely not make archers good if they had not had practice in shooting, neither could it make lancers good, nor horsemen; it cannot even make men able to endure bodily labour, unless they had been trained to it before.

- Cyrus the Great

✧✧✧

D

If the blood of France and of Germany flows again, as it did twenty-five years ago, in a longer and even more murderous war, each of the two peoples will fight with confidence in its own victory, but the most certain victors will be the forces of destruction and barbarism.

- Édouard Daladier, French Premier (1939)

✧✧✧

Good God! This man should be writing dime novels.

- Josephus Daniels, U.S. Secretary of the Navy, regards Billy Mitchell's idea of airplanes sinking a battleship. In July 1921 Mitchell got his experiment and sunk the captured German battleship Ostfreisland. 1921

✧✧✧

Army: A body of men assembled to rectify the mistakes of the diplomats.

- Josephus Daniels

✧✧✧

It is not the strongest of the species that survive, nor the most intelligent, but the one most responsive to change.

- Charles Darwin

✧✧✧

I don't know why I got a medal, all I wanted was to get home alive.

- Shalom David

✧✧✧

I believe that sometimes and especially while young, we need to point ourselves at the heroic. Life most often presents the squalid, the belittling, the prosaic. One's young self does not accept this and the image of the heroic is an anodyne. . . . There was a need to find out, to be probed and proved, to set oneself against the most real of realities, death . . . .

- Patrick Davis, A Child at Arms

✧✧✧

The will to fly was (and in many cases still is) the will to conquer, to overcome all obstacles in the effort to gain control over the natural conditions of environment.

- M.J. Bernard Davy

✧✧✧

I feel not a person but an instrument of destiny

- General Charles De Gaulle, on entering Paris in 1944

✧✧✧

The graveyards are full of indispensable men.

- General Charles De Gaulle

✧✧✧

Tomorrow, the good infantryman will be, no doubt, an accurate marksman—with several types of weapons—but he will also be observer, pioneer signaller, wireless operator, motor car driver, gunner and a camouflage expert.

- General Charles De Gaulle

✧✧✧

Patriotism is when love of your own people comes first; nationalism, when hate for people other than your own comes first.

- General Charles De Gaulle

✧✧✧

A true leader always keeps an element of surprise up his sleeve, which others cannot grasp but which keeps his public excited and breathless.

- General Charles De Gaulle

✧✧✧

Authority doesn't work without prestige, or prestige without distance.

- General Charles De Gaulle

✧✧✧

Deliberation is the work of many men. Action, of one alone.

- General Charles De Gaulle

✧✧✧

Diplomats are useful only in fair weather. As soon as it rains they drown in every drop.

- General Charles De Gaulle

✧✧✧

Don't ask me who's influenced me. A lion is made up of the lambs he's digested, and I've been reading all my life.

- General Charles De Gaulle

✧✧✧

Faced with crisis, the man of character falls back on himself. He imposes his own stamp of action, takes responsibility for it, makes it his own.

- General Charles De Gaulle

✧✧✧

For glory gives herself only to those who have always dreamed of her.

- General Charles De Gaulle

✧✧✧

One cannot govern with 'buts'.

- General Charles De Gaulle

✧✧✧

Politics is too serious a matter to be left to the politicians.

- General Charles De Gaulle

✧✧✧

Since a politician never believes what he says, he is quite surprised to be taken at his word.

- General Charles De Gaulle

✧✧✧

Treaties, you see, are like girls and roses; they last while they last.

- General Charles De Gaulle

✧✧✧

Diplomats are useful only in fair weather. As soon as it rains they drown in every drop.

- General Charles De Gaulle

✧✧✧

Patriotism is when love of your own people comes first; nationalism, when hate for people other than your own comes first.

- General Charles De Gaulle

✧✧✧

France has lost a battle. But France has not lost the war. A makeshift government may have capitulated, giving way to panic, forgetting honour, delivering their country into slavery. Yet nothing is lost! Nothing is lost because this war is a world war. In the free universe, immense forces have not yet been brought into play. Some day these forces will crush the enemy. On that day France must be present at the victory. She will then regain her liberty and her greatness. That is why I ask all Frenchmen, wherever they may be, to unite with me in action, in sacrifice and in hope. Our country is in danger of death. Let us fight to save it.

- General Charles De Gaulle

✧✧✧

Economics is war pursued by other means.

- Raymond F. DeVoe

✧✧✧

Beware lest in your anxiety to avoid war you obtain a master.

- Demosthenes (384-322 B.C.)

✧✧✧

We need money, for sure, Athenians, and without money nothing can be done that ought to be done.

- Demosthenes

✧✧✧

It is impossible to gain permanent power by injustice, perjury, and falsehood.

- Demosthenes

✧✧✧

The man who deems himself born only to his parents will wait for his natural and destined end; the son of his country is willing to die rather than see her enslaved, and will look upon those outrages and indignities, which a commonwealth in subjection is compelled to endure, as more dreadful than death itself.

- Demosthenes

✧✧✧

All speech is vain and empty unless it be accompanied by action.

- Demosthenes

✧✧✧

He who confers a favour should at once forget it, if he is not to show a sordid ungenerous spirit. To remind a man of a kindness conferred and to talk of it, is little different from reproach.

- Demosthenes

✧✧✧

Small opportunities are often the beginning of great enterprises.

- Demosthenes

✧✧✧

Every advantage in the past is judged in the light of the final issue.

- Demosthenes

✧✧✧

You cannot have a proud and chivalrous spirit if your conduct is mean and paltry; for whatever a man's actions are, such must be his spirit.

- Demosthenes

✧✧✧

Minds are like parachutes. They only function when they are open.

- Sir James Dewar

✧✧✧

A person consists of desires, and as is his desire, so is his will; and as is his will, so is his deed; and whatever deed he does, that he will reap.

- Dhammapada

✧✧✧

Hatred will never cease in those who entertain thoughts of revenge.

- Dhammapada

✧✧✧

Wisdom is attained by spiritual insight or intuition rather than by observation and analysis. It is the result of a contemplative rather than an intellectual attitude.

- Dhammapada

✧✧✧

Take courage! Some things you will think for yourself; others God will put into your heart.

- Dhammapada

✧✧✧

If on a journey, the traveller does not meet his better or equal, let him firmly pursue his journey by himself; there is no companionship with a fool.

- Dhammapada

✧✧✧

If a fool be associated with a wise man even all his life, he does not perceive the truth, even as a spoon does not perceive the taste of a dish.

- Dhammapada

✧✧✧

Conquest of self is better than the conquest of others.

- Dhammapada

✧✧✧

Some run swiftly; some walk, some creep painfully; but everyone will reach the goal who keeps on going.

- Dhammapada

✧✧✧

Follow me if I advance! Kill me if I retreat! Revenge me if I die!

- Ngo Dinh Diem
(on becoming President of Vietnam)

✧✧✧

It is a law of nature, common to all mankind, which time shall neither annul nor destroy, that those who have greater strength and power shall bear rule over those who have less.

- Dionysius

✧✧✧

All of our dreams can come true if we have the courage to pursue them.

- Walt Disney

✧✧✧

How much easier it is to be critical than to be correct.

- Benjamin Disraeli

✧✧✧

Next to knowing when to seize an opportunity, the most important thing in life is to know when to forego an advantage.

- Benjamin Disraeli

✧✧✧

There are three kinds of lies: lies, damned lies, and statistics.

- Benjamin Disraeli

✧✧✧

Change is inevitable. Change is constant.

- Benjamin Disraeli

✧✧✧

Characters do not change. Opinions alter, but characters are only developed.

- Benjamin Disraeli

✧✧✧

The services in wartime are fit only for desperadoes, but in peace are only fit for fools.

- Benjamin Disraeli

✧✧✧

Circumstances are beyond human control, but our conduct is in our own power.

- Benjamin Disraeli

✧✧✧

The governments of the present day have to deal not merely with other governments, with emperors, kings and ministers, but also with the secret societies which have everywhere their unscrupulous agents, and can at the last moment upset all the governments' plans.

- Benjamin Disraeli

✧✧✧

The world is governed by very different personages from what is imagined by those who are not behind the scenes.

- Benjamin Disraeli

✧✧✧

Where knowledge ends, religion begins.

- Benjamin Disraeli

✧✧✧

Old men are more cautious than young men, and less able to make quick decisions than those whose arteries have not begun to harden.

- Norman Dixon
On The Psychology of Military Incompetence

✧✧✧

One of the chief causes and characteristics of incompetence is an aversion to any pursuit that smacked of intellectualism.

- Norman Dixon
On The Psychology of Military Incompetence

✧✧✧

I have concluded that some British generals were bunglers and butchers. The two went together because, under the conditions of warfare, butchery was the result of bungling... The more senior the rank, the greater the influence and authority—and much greater the responsibility.

- Norman Dixon
On The Psychology of Military Incompetence

✧✧✧

Smith-Dorrien being told by French to 'do as you are ordered and don't ask questions' when, having been told to give battle on the line of the Mons canal, he had asked: 'Do you mean to take the offensive or stand on the defensive?'

- Norman Dixon
On The Psychology of Military Incompetence

✧✧✧

Only the most blinkered could deny that the First World War exemplified every aspect of high-level military incompetence. For sheer lack of imaginative leadership, inept decisions, ignoring the military intelligence, underestimation of the enemy, delusional optimism and monumental wastage of human resources it has surely never had its equal'

- Norman Dixon
On The Psychology of Military Incompetence

✧✧✧

The issue is not war and peace, rather, how best to preserve our freedom.

- General Russell E. Dougherty (1920–2007)
Chief of the US Strategic Air Command

✧✧✧

To conquer the command of the air means victory; to be beaten in the air means defeat.

- Giulio Douhet, The Command of the Air, 1921

✧✧✧

To have command of the air means to be able to cut an enemy's army and navy off from their bases of operation and nullify their chances of winning the war.

- General Giulio Douhet

✧✧✧

In order to assure an adequate national defence, it is necessary—and sufficient—to be in a position in case of war to conquer the command of the air.

- General Giulio Douhet

✧✧✧

And even if a semblance of order could be maintained and some work done, would not the sight of a single enemy airplane be enough to induce a formidable panic? Normal life would be unable to continue under the constant threat of death and imminent destruction.

- General Giulio Douhet, 'Il dominio dell'aria,' 1921.

✧✧✧

Because of its independence of surface limitations and its superior speed the airplane is the offensive weapon par excellence.

- General Giulio Douhet

✧✧✧

I have a mathematical certainty that the future will confirm my assertion that aerial warfare will be the most important element in future wars, and that in consequence not only will the importance of the Independent Air Force rapidly increase, but the importance of the army and navy will decrease in proportion.

- General Giulio Douhet, 'Command of the Air,' 1921.

✧✧✧

Aerial warfare will be the most important element in future wars.

- Giulio Douhet

✧✧✧

Would not the sight of a single enemy airplane be enough to induce a formidable panic? Normal life would be unable to continue under the constant threat of death and imminent destruction.

- General Giulio Douhet, 'The Command of the Air,' 1921.

✧✧✧

The battlefield in the air will be the decisive one.

- Douhet

✧✧✧

In order to assure an adequate national defence, it is necessary—and sufficient—to be in a position in case of war to conquer the command of the air.

- General Giulio Douhet

✧✧✧

In order not to be left behind, we must work fast-clay feet are irreconcilable with the lightness of wings.

- Giulio Douhet (Command of the Air)

✧✧✧

All (successful commanders) maintained discipline at an equal standard. All fired their soldiers to the utmost pitch in battle, all encouraged them to bear privation in the field, and bore it with them. All equally won their soldiers' hearts. All obtained this control over men by scrupulous care of their army's welfare, courage equal to any test, readiness to participate in the heat and labour of the day, personal magnetism, justice in rewards and punishments, friendliness in personal intercourse, and power of convincing men.

- Theodore A. Dodge (Great Captains)

✧✧✧

The enemy holds every trump card, covering all areas with long-range air patrols and using location methods against which we still have no warning... The enemy knows all our secrets and we know none of his.

- Grand Admiral Doenitz
Commander in Chief of the German Navy (1943)

✧✧✧

The country will someday pay for the stupidities of those who were in the majority on this commission. They know as much about the future of aviation as they do about the sign writing of the Aztecs.

- General James H. Doolittle, at the time an Air Corps Reservist and racing pilot, comments to a reporter regards the presidential committee of inquiry that did not support establishment of an air force separate from the army, 1934.

✧✧✧

Integrity; A man's word is his bond.

- General Jimmy Doolittle
U.S Army Air Forces, leader of the WWII Tokyo Raid

✧✧✧

The function of the Army and Navy in any future war will be to support the dominant air arm.

- General James H. Doolittle, US Air Force (1896–1993)
Georgetown University, 1949.

✧✧✧

Adolf Galland said that the day we took our fighters off the bombers and put them against the German fighters, that is, went from defensive to offensive, Germany lost the air war. I made that decision and it was my most important decision during World War II. As you can imagine, the bomber crews were upset. The fighter pilots were ecstatic.

- General James H. Doolittle

✧✧✧

"He who wants to protect everything, protects nothing," is one of the fundamental rules of defence.

- General James H. Doolittle

✧✧✧

In order to gauge the level of discipline within a Unit, the military developed eight indicators. Like many of its checklists, this list of indicators is uncomplicated and leaves a broad latitude for interpretation.

- All missions are accomplished.
- Soldiers have confidence and a sharp appearance.
- Soldiers are proud of their unit; they know it has a good reputation.
- Weapons and equipment are well-maintained.
- Soldiers at all levels are actively engaged in doing their duties. They do not waste time.
- Soldiers cooperate; they willingly help one another.
- Training is well planned, well conducted, and thoroughly evaluated for individual and unit strengths and weaknesses.
- The unit fights successfully under stress.
- An experienced eye can detect the measure of discipline within a unit by merely observing something as simple as a soldier's hand salute. A sharp, crisp greeting reflects personal and organizational pride and a positive feeling that proclaims confidence, professionalism, and a "can-do" approach to all tasks. From this utterly routine gesture, his entire unit's morale, combat readiness, and leadership can be surmised without setting foot on its premises.

- Gil Dorland and John Dorland, Duty, Honour, Company
West Point Fundamentals for Business Success, 1992

✧✧✧

The [Positive Command Climate] programme ... details specific actions a leader can take to build a positive command climate. These concepts, noted in Manual of Common Tasks, are not altogether new; they are very similar to the leadership fundamentals that were ingrained in us as cadets.

- Communicate a sense of vision or focus.
- Maintain a proper focus in all training activities.
- Establish high, attainable, clearly understood standards.
- Encourage competition against standards rather than each other.
- Allow subordinates the freedom to exercise initiative.
- Establish accountability at the proper level.
- Show confidence in subordinates.
- Encourage and reward prudent risk taking.
- Achieve high performance through positive motivation and rewards.
- Underwrite honest mistakes.
- Share decision making with subordinates when appropriate.
- Give clear missions and indicate where subordinates have discretion and where they do not.
- Listen to your subordinates and seek their ideas.
- Demonstrate concern about the welfare of subordinates.
- Establish and model high standards.
- Practice what you preach.

- Gil and John Dorland,
Duty, Honour, Company, West Point Fundamentals for Business Success, 1992

✧✧✧

I asked for a few Americans. They brought with them the courage of a whole army.

- General Abdul Rashid Dostum
on the Green Berets in November 2001

✧✧✧

The best defence of the country is the fear of the fighter. If we are strong in fighters we should probably never be attacked in force. If we are moderately strong we shall probably be attacked and the attacks will gradually be bought to a standstill. . . . If we are weak in fighter strength, the attacks will not be bought to a standstill and the productive capacity of the country will be virtually destroyed.

- Air Chief Marshal Sir Hugh Dowding

✧✧✧

In the early stages of the fight Mr. Winston Churchill spoke with affectionate raillery of me and my "Chicks." He could have said nothing to make me more proud; every Chick was needed before the end.

- Air Chief Marshal Sir Hugh C. T. Dowding, dispatch to the Secretary of State for Air, 20 August 1941.

✧✧✧

Said the king to the colonel,
'The complaints are eternal,
That you Irish give more trouble
Than any other corps.

Said the colonel to the king,
'This complaint is no new thing,
For your foemen, sire, have made it
A hundred times before.

- Sir Arthur Conan Doyle (The Irish Colonel)

✧✧✧

How often have I said to you that when you have eliminated the impossible, whatever remains, however improbable, must be the truth?

- Sir Arthur Conan Doyle
Sherlock Holmes [in The Sign of Four, 1890]

✧✧✧

A decision should always be made at the lowest possible level and as close to the scene of action as possible. However, a decision should always be made at a level insuring that all activities and objectives are fully coordinated.

- Peter F. Drucker

✧✧✧

There is nothing so useless as doing efficiently that which should not be done at all.

- Peter Drucker

✧✧✧

My greatest strength as a consultant is to be ignorant and ask a few questions.

- Peter F. Drucker

✧✧✧

Checking the results of a decision against its expectations shows executives what their strengths are, where they need to improve, and where they lack knowledge or information.

- Peter F. Drucker

✧✧✧

Effective leadership is not about making speeches or being liked; leadership is defined by results not attributes.

- Peter F. Drucker

✧✧✧

Efficiency is doing things right; effectiveness is doing the right things.

- Peter F. Drucker

✧✧✧

Executives owe it to the organization and to their fellow workers not to tolerate nonperforming individuals in important jobs.

- Peter F. Drucker

✧✧✧

Most discussions of decision making assume that only senior executives make decisions or that only senior executives' decisions matter. This is a dangerous mistake.

- Peter F. Drucker

✧✧✧

Trying to predict the future is like trying to drive down a country road at night with no lights while looking out the back window.

- Peter F. Drucker

✧✧✧

The world will never have lasting peace so long as men reserve for war the finest human qualities. Peace, no less than war, requires idealism and self-sacrifice and a righteous and dynamic faith.

- John Foster Dulles

✧✧✧

Through most of its wars, the United States successfully used the attrition approach. It is easier to be proficient at this type of warfare. You need to master only the simplest military skills and possess enormous quantities of arms and munitions.

- James F. Dunnigan (How To Make War, 1993)

✧✧✧

Although the story sounds apocryphal, it was told by a Russian tank officer who claimed to have been a witness. A tank unit was out on manoeuvres that involved defence against enemy attack helicopters. The exercise proved a frustrating one for the tankers. They were unable to evade detection, and, worse, the umpires constantly ruled that the simulated shots that they were taking at the target drones-the army budget for such being tight-were misses. Finally, one sergeant boiled over. Unbeknown to his superior, he took careful aim and cut loose with live ammunition, scoring a perfect hit and bringing the drone crashing down. The ensuing bureaucratic furore was resolved in perfect bureaucratic fashion. The powers-that-were decided to reprimand the platoon commander for the poor discipline in his outfit, but recognized the sergeant's marksmanship with a commendation.

- James F. Dunnigan and Albert A. Nofi, Dirty Little Secrets; Military Information You're Not Supposed to Know, 1990

✧✧✧

The individual soldier has become increasingly independent in combat; this in turn has not only called for improved training, discipline, motivation and coordination, it has also required fostering improvement in intelligence, initiative, and judgement on the part of each individual at lower levels.

- Colonel T.N. Dupuy, US Army
The Evolution of Weapons and Warfare, 1980

✧✧✧

My personal feeling is that if I have done anything worthwhile, it is in military theory and the relationship of the elements of historical experience to theory.

- Colonel T.N. Dupuy, US Army
The Evolution of Weapons and Warfare, 1980

✧✧✧

I was brought up by my father to be both a soldier and a military historian. To him the two were inseparable, and that is the way it has always been for me.

- Colonel T.N. Dupuy, US Army
The Evolution of Weapons and Warfare, 1980

✧✧✧

The ratio of the tank loss rate to the casualty rate appears to be a function of the density or proportion of tanks in the force. A force can be considered armour-heavy when the proportion of tanks exceeds 6 per 1,000 troops. When two armour-heavy forces engage in battle, whether for a short or long period, the ratio of tank loss rates to the casualty rates will be in the range of 5.00 to 8.00 for the successful force (with an average value of about 6.00) & much higher for the losing force. There are, of course, exceptions, as there always are in combat data, where each event is unique, but the pattern is clear. Furthermore the ratio of tank loss rates to casualty rates appears to remain constant as the proportion of tanks increases above 6 per 1,000 troops. On the other hand, that ratio decreases, apparently in more-or-less linear fashion, when the proportion of tanks decline below 6 per 1,000 troops. It appears further that the range of the ratio for an armour-heavy force is likely to be between about 5 & 7 for a winner & between about 7 and 10 (sometimes much higher) for a loser.

- Colonel T.N. Dupuy, US Army
The Evolution of Weapons and Warfare, 1980

✧✧✧

Despite differences of opinion as to the relevance of military history, and its significance as a guide to what happens or is likely to happen in modern combat, it is very common to hear strategic concepts or theories presented with the introductory words: "Military history proves that ......"

- Colonel T.N. Dupuy, US Army
Numbers, Predictions and War

✧✧✧

Military history is relevant to modern combat, as long as due allowances are made for the significant (but not usually overwhelming) impact of changing weapons and technology—particularly in the fields of communications, observation, and transportation.

- Colonel T.N. Dupuy, US Army
Numbers, Predictions and War

✧✧✧

E

[In World War I] air raids on both sides caused interruptions to production and transportation out of all proportion to the weight of bombs dropped.

- Edward Mead Earle
Makers of Modern Strategy

✧✧✧

History teaches us that men and nations behave wisely once they have exhausted all other alternatives.

- Abba Eban (1915-2002)
Israeli diplomat and politician.

✧✧✧

Arabs never miss an opportunity to miss an opportunity (i.e., for peace).

- Abba Eban

✧✧✧

History teaches us that men and nations behave wisely once they have exhausted all other alternatives.

- Abba Eban, *Speech in London (16 December 1970)*

✧✧✧

I think that this is the first war in history that on the morrow the victors sued for peace and the vanquished called for unconditional surrender.

- Abba Eban "The Impact of the Six-Day War"

✧✧✧

A consensus means that everyone agrees to say collectively what no one believes individually.

- Abba Eban

✧✧✧

Finally, as you grow older, try not to be afraid of new ideas. New or original ideas can be bad as well as good, but whereas an intelligent man with an open mind can demolish a bad idea by reasoned argument, those who allow their brains to atrophy resort to meaningless catchphrases, to derision and finally to anger in the face of anything new.

- The Duke of Edinburgh, speaking at the Sovereign's parade, R.M.C., 1955, quoted in On the Psychology of Military Incompetence, Norman F. Dixon

✧✧✧

There will one day spring from the brain of science a machine or force so fearful in its potentialities, so absolutely terrifying, that even man, the fighter, who will dare torture and death in order to inflict torture and death, will be appalled, and so abandon war forever.

- Thomas A. Edison (1847-1931)

✧✧✧

I have not failed. I've just found 10,000 ways that won't work.

- Thomas A. Edison

✧✧✧

The value of an idea lies in the using of it.

- Thomas A. Edison

✧✧✧

There is far more opportunity than there is ability.

- Thomas A. Edison

✧✧✧

All's fair in love and war.

- Francis Edwards

✧✧✧

I know not with what weapons World War III will be fought, but World War IV will be fought with sticks and stones.

- Albert Einstein (contemplating nuclear devastation)

✧✧✧

No problem can be solved from the same consciousness that created it. We must learn to see the world anew.

-Albert Einstein

✧✧✧

The unleashed power of the atom has changed everything except our modes of thinking, and we thus drift toward unparalleled catastrophes. The solution to this problem lies in the heart of mankind. If only I had known, I should have become a watchmaker.

- Albert Einstein

✧✧✧

When I have one week to solve a seemingly impossible problem, I spend six days defining the problem. Then, the solution becomes obvious.

- Albert Einstein

✧✧✧

The time we live in requires a new way of thinking.

- Albert Einstein

✧✧✧

Any intelligent fool can make things bigger and more complex. It takes a touch of genius and a lot of courage to move in the opposite direction.

- Albert Einstein

✧✧✧

One cannot simultaneously prevent and prepare for war

- Albert Einstein

✧✧✧

Force always attracts men of low morality.

- Albert Einstein

✧✧✧

I made one great mistake in my life-when I signed the letter to President Roosevelt recommending that atom bombs be made but there was some justification-the danger that the Germans would make them.

- Albert Einstein

✧✧✧

Nationalism is an infantile disease. It is the measles of mankind.

- Albert Einstein

✧✧✧

Peace cannot be kept by force; it can only be achieved by understanding.

- Albert Einstein

✧✧✧

It is characteristic of the military mentality that nonhuman factors (atom bombs, strategic bases, weapons of all sorts, the possession of raw materials, etc.) are held essential, while the human being, his desire and thoughts—in short, the psychological factors—are considered as unimportant and secondary...The individual is degraded...to 'human material.'

- Albert Einstein

✧✧✧

Through the release of atomic energy, our generation has brought into the world the most revolutionary force since prehistoric man's discovery of fire. This basic force of the universe cannot be fitted into the outmoded concept of narrow nationalisms. For there is no secret and there is no defence; there is no possibility of control except through the aroused understanding and insistence of the peoples of the world. We scientists recognize our inescapable responsibility to carry to our fellow citizens an understanding of atomic energy and its implication for society. In this lies our only security and our only hope—we believe that an informed citizenry will act for life and not for death.

- Albert Einstein

✧✧✧

The release of atomic energy has not created a new problem. It has merely made more urgent the necessity of solving an existing one.

- Albert Einstein

✧✧✧

It may affront the military-minded person to suggest a regime that does not maintain any military secrets.

- Albert Einstein

✧✧✧

He who joyfully marches to music in rank and file has already earned my contempt. He has been given a large brain by mistake, since for him the spinal cord would fully suffice. This disgrace to civilization should be done away with at once. Heroism at command, senseless brutality, and all the loathsome nonsense that goes by the name of patriotism, how violently I hate all this, how despicable and ignoble war is; I would rather be torn to shreds than be part of so base an action! It is my conviction that killing under the cloak of war is nothing but an act of murder.

- Albert Einstein

✧✧✧

When you put on a uniform there are certain inhibitions that you accept.

- General Dwight D. Eisenhower (34th President of USA (1953-1961)

✧✧✧

Every gun that is made, every warship launched, every rocket fired, signifies in the final sense a theft from those who hunger and are not fed, those who are cold and are not clothed.

- General Dwight D. Eisenhower

✧✧✧

From behind the Iron Curtain, there are signs that tyranny is in trouble and reminders that its structure is as brittle as its surface is hard.

- General Dwight D. Eisenhower

✧✧✧

History does not long entrust the care of freedom to the weak or the timid.

- General Dwight D. Eisenhower

✧✧✧

What counts is not necessarily the size of the dog in the fight—it's the size of the fight in the dog.

- General Dwight D. Eisenhower

✧✧✧

Military operations are drastically affected by many considerations, one of the most important of which is the geography of the region.

- General Dwight D. Eisenhower

✧✧✧

Every gun that is made, every warship launched, every rocket fired signifies in the final sense, a theft from those who hunger and are not fed, those who are cold and are not clothed. This world in arms is not spending money alone. It is spending the sweat of its labourers, the genius of its scientists, the hopes of its children. This is not a way of life at all in any true sense. Under the clouds of war, it is humanity hanging on a cross of iron.

- Dwight D. Eisenhower,
speech to American Society of Newspaper Editors, 16 April 1953

✧✧✧

The problem in defence is how far you can go without destroying from within what you are trying to defend from without.

- General Dwight D. Eisenhower

✧✧✧

I hate war as only a soldier who has lived it can, only as one who has seen its brutality, its futility, its stupidity.

- General Dwight D. Eisenhower

✧✧✧

I deplore the need or the use of troops anywhere to get American citizens to obey the orders of constituted courts.

- General Dwight D. Eisenhower

✧✧✧

If men can develop weapons that are so terrifying as to make the thought of global war include almost a sentence for suicide, you would think that man's intelligence and his comprehension... would include also his ability to find a peaceful solution.

- General Dwight D. Eisenhower

✧✧✧

If the United Nations once admits that international disputes can be settled by using force, then we will have destroyed the foundation of the organization and our best hope of establishing a world order.

- General Dwight D. Eisenhower

✧✧✧

Leadership is the art of getting someone else to do something you want done because he wants to do it.

- General Dwight D. Eisenhower

✧✧✧

Speeches are for the younger men who are going places. And I'm not going anyplace except six feet under the floor of that little chapel adjoining the museum and library at Abilene.

- General Dwight D. Eisenhower

✧✧✧

The sergeant is the Army.

- General Dwight D. Eisenhower

✧✧✧

There is no victory at bargain basement prices.

- General Dwight D. Eisenhower

✧✧✧

In this war, which was total in every sense of the word, we have seen many great changes in military science. It seems to me that not the least of these was the development of psychological warfare as a specific and effective weapon.

- General Dwight D. Eisenhower

✧✧✧

You don't lead by hitting people over the head-that's assault, not leadership.

- General Dwight D. Eisenhower

✧✧✧

The eyes of the world are upon you. The hopes and prayers of liberty-loving people everywhere march with you.

- General Dwight D Eisenhower, address to his troops D-Day, 1944

✧✧✧

Without a doubt, psychological warfare has proven its right to a place of dignity in our military arsenal.

- General Dwight D. Eisenhower

✧✧✧

Every war is going to astonish you in the way it has occurred and in the way it is carried.

- President Dwight D. Eisenhower

✧✧✧

The Normandy Invasion was based on a deep-seated faith in the power of the air forces, in overwhelming numbers, to intervene in the land battle.... Without that air force, without the aid of the enemy air force out of the sky, without its power to intervene in the land battle, that invasion would have been fantastic.... Unless we had faith in the air power to intervene and make safe that landing, it would have been more than fantastic, it would have been criminal.

- Gen Dwight D. Eisenhower, 1945

✧✧✧

Though force can protect in emergency, only justice, fairness, consideration and cooperation can finally lead men to the dawn of eternal peace.

- General Dwight D. Eisenhower

✧✧✧

The conjunction of an immense military establishment and a huge arms industry is new in the American experience. The total influence—economic, political, and even spiritual—is felt in every city, every state house, and every office of the federal government... In the councils of government, we must guard against the acquisition of unwarranted influence, whether sought or unsought, by the military-industrial complex.

- General Dwight D. Eisenhower

✧✧✧

You have a row of dominoes set up; you knock over the first one, and what will happen to the last one is that it will go over very quickly.

- General D. Eisenhower, 1954

✧✧✧

Only those who risk going too far can possibly find out how far one can go.

- T.S. Eliot

✧✧✧

We shall not cease from exploring
And at the end of our exploration
We will return to where we started
And know the place for the first time.
Now that's in a sense where I'm beginning to be.

- T.S. Eliot

✧✧✧

There is nothing that war has ever achieved we could not better achieve without it.

- Havelock Ellis

✧✧✧

Another morale booster among officers in an Infantry battalion in Korea was the "Muddler" award. The award was an enlarged drink stirring rod about the size of a canoe paddle. It was painted black and appropriately decorated with a battalion crest. The award was presented to the officer in the battalion who muddled or bungled the most in the performance of his duties during a two week period. The previous winner presented the "Muddler" to a worthy successor. he received nominations from the officers present and then selected the officer he deemed the biggest muddler of them all. There were some real bungles reported. One such bungle involved the battalion's reconnaissance platoon leader who, while attempting to find a fording site in a swollen river, found himself standing in ankle deep water on the hood of his jeep. The "Muddler" award stimulated laughter and contributed to the esprit de corps of the battalion. It also encouraged zero defects.

- Captain Peter M. Elson
"Humour is no Laughing Matter", Infantry, March-April 1972

✧✧✧

As soon as technological advances may be applied to military goals, and furthermore are already used for military purposes, they almost immediately seem obligatory, and also often go against the will of the commanders in triggering changes or even revolutions in the modes of combat.

- Frederich Engels

As explained by Liddell Hart, the revived infantryman would be 'tria juncta in uno -stalker, athlete, and marksman.' Equipped with a lighter rifle, to alleviate ammunition supply, and trained to a high standard in field craft, he would be capable of destroying enemy machine- gun and antitank positions through stealth and deadly accurate small-arms fire. Unlike his Great War counterpart, he would not be a beast of burden carrying 70 pounds of personal kit; rather, he would carry but on-third of his own weight. Dressed as an athlete and 'light of foot,' he would also be 'quick of thought' and capable of acting on his own or as part of an independent team. The elastic chain of little groups would replace the traditional infantry line.

- John A. English, A Perspective on Infantry

✧✧✧

By prompt and imaginative action, lone riflemen and companies sometimes diverted whole enemy corps, while a machine-gun squad at a roadblock began the defeat of an armoured division. In short, though mass was there somewhere in support, many great victories pivoted upon the fire action of a very few.

- John A. English, A Perspective on Infantry

✧✧✧

A German attacking force, even in the smallest details of combat, maintained in this manner superiority, initiative, and surprise. This concept of operations has been referred to by some commentators as the principle of the 'unlimited objective.

- John A. English, A Perspective on Infantry

✧✧✧

His Majesty made you a major because he believed you would know when not to obey orders." ... As far as the Germans were concerned, the first demand in war was decisive action.

- John A. English, A Perspective on Infantry, 1981

✧✧✧

To exploit the foot soldier's loco-mobility to maximum advantage, infantry units should not be fettered by having to adapt their formations rigidly to the movements of tanks or artillery barrages. If they are thus restricted, they are apt to lose both their initiative and special value in battle.

- John A. English, A Perspective on Infantry

✧✧✧

Freedom is not procured by a full enjoyment of what is desired, but by controlling the desire.

- Epictetus, Roman philosopher (50–138 AD)

✧✧✧

It is not death or pain that is to be dreaded, but the fear of pain or death.

- Epictetus, Roman philosopher (50–138 AD)

✧✧✧

Death does not concern us, because as long as we exist, death is not here. And when it does come, we no longer exist.

- Epicurus

✧✧✧

War is delightful to those who have had no experience of it.

- Desiderius Erasmus

✧✧✧

A good portion of speaking well consists in knowing how to lie.

- Desiderius Erasmus

✧✧✧

The most disadvantageous peace is better than the most just war.

- Deciderius Erasmus

✧✧✧

People see only what they want to see.

- Ralph Waldo Emerson

✧✧✧

The characteristic of a genuine heroism is its persistency. All men have wandering impulses, fits and starts of generosity. But when you have resolved to be great, abide by yourself, and do not weakly try to reconcile yourself with the world. The heroic cannot be common, nor the common heroic.

- Ralph Waldo Emerson

✧✧✧

Whatever you do, you need courage. Whatever course you decide upon, there is always someone to tell you that you are wrong. There are always difficulties arising that tempt you to believe your critics are right.

- Ralph Waldo Emerson

✧✧✧

A great part of courage is the courage of having done the thing before.

- Ralph Waldo Emerson

✧✧✧

When duty whispers low, 'thou must', the youth replies, 'I can!'

- Ralph Waldo Emerson

✧✧✧

The strong take from the weak, but the smart take from the strong.

- Ralph Waldo Emerson

✧✧✧

Leadership is the courage to admit mistakes, the vision to welcome change, the enthusiasm to motivate others, and the confidence to stay out of step when everyone else is marching to the wrong tune.

- E. M. Estes

✧✧✧

The God of War hates those who hesitate.

- Euripides, 480-406 B.C.

✧✧✧

Courage may be taught as a child is taught to speak.

- Euripides

✧✧✧

The God of War hates those who hesitate.

- Euripides, 480-406 B.C. Greek tragic dramatist

✧✧✧

Education is a better safeguard of liberty than a standing army.

- Edward Everett

✧✧✧

There is nothing impossible! Give your orders, support them with firmness, and you will see every obstacle vanish!

- Lieutenant-General Johann von Ewald

✧✧✧

The injustice of defeat lies in the fact that its most innocent victims are made to look like heartless accomplices. It is impossible to see behind defeat, the sacrifices, the austere performance of duty, the self-discipline and the vigilance that are there—those things the god of battle does not take account of.

- Antoine De Saint-Exupery

✧✧✧

F

... There remains only France .... If we succeeded in opening the eyes of her people to the fact that in a military sense they have nothing more to hope for, that breaking point would be reached and England's best sword knocked out of her hand. To achieve that object the uncertain method of a mass breakthrough, in any case beyond our means, is unnecessary. We can probably do enough for our purposes with limited resources. Within our reach behind the French sector of the Western front there are objectives for the retention of which the French General Staff would be compelled to throw in every man they have. If they do so the forces of France will bleed to death.

- Erich von Falkenhayn
German chief of staff, in a Christmas, 1915 letter to the Kaiser

✧✧✧

So long as large armies go to battle, so long will the air arm remain their spearhead.

- Cyril Falls (The Nature of Modern War)

✧✧✧

Those who study warfare only in the light of history think of the next war in terms of the last. But those who neglect history deprive themselves of a yardstick by which theory can be measured.

- Cyril Falls (The Nature of Modern War)

✧✧✧

Damn the torpedoes, Full speed ahead!

- Admiral David Glasgow Farragut (1801-1870).
Aboard Hartford, Farragut entered Mobile Bay, Alabama, 5 August 1864, in two columns, with armoured monitors leading and a fleet of wooden ships following. When the lead monitor Tecumseh was demolished by a mine, the wooden ship Brooklyn stopped, and the line drifted in confusion toward Fort Morgan. As disaster seemed imminent, Farragut gave the orders embodied by these famous words. He swung his own ship clear and headed across the mines, which failed to explode. The fleet followed and anchored above the forts, which, now isolated, surrendered one by one. The torpedoes to which Farragut and his contemporaries referred would today be described as tethered mines

✧✧✧

The effects could well be called unprecedented, magnificent, beautiful, stupendous, and terrifying. No man-made phenomenon of such tremendous power had ever occurred before ...... Thirty seconds after the explosion came first, the air blast pressing hard against people and things, to be followed almost immediately by the strong, sustained awesome roar which warned of doomsday and made us feel that we puny things were blasphemous to dare tamper with the forces heretofore reserved to the Almighty.

- General Thomas Farrell
Report on America's successful atomic bomb test at Alamogordo, New Mexico, 1945

✧✧✧

The Regiment is mother, sister and mistress.... It is a high, an admirable phase of patriotism, for, to the soldier, his regiment is his country.' Men died for the honour of their regiment. 'Forward the 53rd!' was a more potent cry than 'Forward for Britain!'

- Byron Farwell, Queen Victoria's Little Wars, 1972

✧✧✧

Winston Churchill on Saluting:

> On the 4th September 1940, Mr. Winston Churchill visited and inspected units of the 2nd AIF then encamped on Salisbury Plain. While passing down the ranks of the writer's battalion, the Prime Minister keenly scrutinized the men, meanwhile asking a number of questions on the state of training, supply of unit equipment and so forth. As is well known, Mr. Churchill was first commissioned in the 4th Hussars and, during the Great War of 1914-18, at one period, he commanded the 6th Royal Scots Fusiliers. It was clear that his own regimental training, his possible association with men of the 1st AIF in France, and speculation on the qualities of the new Anzacs, inspired the final question in this interrogation: "How are they on saluting?" The answer to this was followed by one of those inimitable comments which, like so many of the famous statesman's utterances gets down to the roots of the matter in arresting phraseology. He said: "You know, in my young subaltern days, I was always taught that saluting was the outward and visible sign of an inward and spiritual grace."

In May 1944, when the United Kingdom was crammed with British and American troops in training for D Day, a questioner in the House of Commons asked the Prime Minister if he would consider an order that would eliminate the obligation to salute when off duty. Mr. Churchill's reply is quoted in full: "No Sir: a salute is an acknowledgement of the King's Commission and a courtesy to Allied Officers, and I do not consider it desirable to attempt to make the distinction suggested. If my honourable friend had an opportunity during the war of visiting Moscow he would find the smartest saluting in the world. The importance attached to these minor acts of ceremony builds up armies which are capable of facing the greatest rigours of war."

- Brigadier J. Field, CBE, DSO, ED, 4th Infantry Brigade

✧✧✧

Don't bother just to be better than your contemporaries or predecessors. Try to be better than yourself.

- William Faulkner

✧✧✧

Things to remember:

1. The worth of character;
2. The improvement of talent;
3. The influence of example;
4. The joy of origination;
5. The dignity of simplicity;
6. The success of perseverance;
7. The value of time;
8. The pleasure of working;
9. The obligation of duty;
10. The power of kindness;
11. The wisdom of economy;
12. The virtue of patience.

- Marshall Field

✧✧✧

The essence of war is violence. Moderation in war is imbecility.

- Admiral Lord John Fisher (1841-1920)

✧✧✧

All nations want peace, but they want a peace that suits them.

- Admiral Lord John Fisher

✧✧✧

Any fool can obey orders.

- Admiral Lord John Fisher

✧✧✧

By land and by sea the approaching aircraft development knocks out the present fleet, makes invasion practicable, cancels our country being an island, transforms the atmosphere into a battle ground of the future. There is only thing to do to the ostriches who are spending these vast millions on what is as useful for the next war as bows and arrows. Sack the lot. As the locusts swarmed over Egypt, so will aircraft swarm in the heavens, carrying inconceivable cargoes of men and bombs, some fast and some slow. Some will act like battle cruisers and others as destroyers. All cheap and-this is the gist of it-requiring only a few men as crew.

- Admiral Lord Jack Fisher

✧✧✧

The more highly adapted an organism becomes, the less adaptable it is to any new change.

- R A Fisher (Fisher's Fundamental Theorem)

✧✧✧

A man is a critic when he cannot be an artist, in the same way that a man becomes an informer when he cannot be a soldier.

- Gustave Flaubert

✧✧✧

This delicately poised relationship between military and political interests forms a pattern which recurs throughout the history of British rule in India. The two interests are seldom compatible. The outlook of a political agent and a commander in the field are almost bound to differ, and to demarcate their respective spheres of responsibility in unforeseeable contingencies is impossible. Their partnership is full of the seeds of conflict. Where, in the decision to arrest a tribal leader, to bombard a fort, or to burn down a village, is the border-line between political and military action? Who should have the say when it comes to taking a calculated risk—the man who is carrying out a political directive, or the man who is worrying about the safety of his troops? Friction, or worse, is all but inevitable when control is shared between the military and civil power in a difficult venture. The only hope of averting it is to ensure that each is represented by an individual who is likely to get along well with his opposite number.

- Peter Fleming (1907-1971), Bayonets to Lhasa

✧✧✧

To be disciplined does not mean that one does not commit any break of discipline;...does not mean being silent, abstaining, or doing only what one thinks one may undertake without risk; it is not the art of eluding responsibility; it means acting in compliance with orders received and therefore of finding in one's own mind by effort and reflection the possibility of carrying out such orders. It also means finding in one's own will the energy to face the risks involved in execution.

- Marshal Ferdinand Foch

Airplanes are interesting toys but of no military value.

✧✧✧

- Marshal Ferdinand Foch Professor of Strategy, Ecole Superieure de Guerre (circa 1911)

✧✧✧

The potentialities of aircraft attacks on a large scale are almost incalculable, but it is clear that such attack, owing to its crushing moral effect on a Nation, may impress the public opinion to a point of disarming the Government and thus becoming decisive.

- Marshal Ferdinand Foch, 1922

✧✧✧

The common theory that, in order to win, an army must have superiority of rifles and cannon, better bases, more wisely chosen positions, is radically false. For it leaves out of account the most important part of the problem, that which animates it and makes it live, man—with his moral, physical and intellectual qualities.

- Marshal Ferdinand Foch

✧✧✧

The military mind always imagines that the next war will be on the same lines as the last. That has never been the case and never will be. One of the great factors on the next war will be aircraft obviously. The potentialities of aircraft attack on a large scale are almost incalculable.

- Marshal Ferdinand Foch

✧✧✧

It take 15,000 casualties to train a major general.

- Marshal Ferdinand Foch

✧✧✧

A battle won is a battle in which one will not confess oneself beaten.

- Marshal Ferdinand Foch
The Principles of War

✧✧✧

It is not so necessary to annihilate the enemy combatants as to annihilate their courage. Victory is ours as soon as the enemy has been brought to believe that his cause is lost...an enemy is not to be reduced to impotence by means of complete individual annihilation, but by destroying his hope in victory.

- Marshal Ferdinand Foch
The Principles of War

✧✧✧

The result was secured not by physical means—these were all to the advantage of the vanquished—it was achieved by a purely moral action which alone brought about a decision and a complete decision.

- Marshal Ferdinand Foch
The Principles of War

✧✧✧

Hard pressed on my right. My centre is yielding. Impossible to manoeuvre. Situation excellent. I attack.

- Marshall Ferdinand Foch (1851-1955)
Supreme Commander of Allied Forces in 1918
First Battle of the Marne

✧✧✧

An absence of the "will to conquer" on the commander's part was fatal, regardless of the bravery of the men underneath him. Foch notes that the battles around Metz during August 1870 (Mars-la-Tour, Gravelotte and Noiseville), show "an army fighting bravely but a chief not desiring to secure victory".

- Marshal Ferdinand Foch
The Principles of War

✧✧✧

The most powerful weapon on earth is the human soul on fire.

- Marshal Ferdinand Foch

✧✧✧

The fundamental qualities for good execution of a plan is first; intelligence; then discernment and judgment, which enable one to recognize the best method as to attain it; the singleness of purpose; and, lastly, what is most essential of all, will-stubborn will.

- Marshal Ferdinand Foch

✧✧✧

The power to command has never meant the power to remain mysterious.

- Marshal Ferdinand Foch

✧✧✧

We're entrusted with the security of our nation. The tools of our trade are lethal, and we engage in operations that involve risk to human life and untold national treasure. Because of what we do our standards must be higher than those of society at large.

- General Ronald R. Fogleman

✧✧✧

The tragedy of war is that it uses man's best to do man's worst.

- Henry Fosdick

✧✧✧

If you think you can, or if you think you can't-either way you're right!

- Henry Ford

✧✧✧

Praise the Lord and pass the ammunition!

- Lieutenant Howell Maurice Forgy, US Navy chaplain serving in the heavy cruiser USS New Orleans during the Japanese attack on Pearl Harbour on 7 December 1941, saw the men of an ammunition party tiring as they laboured to bring shells to the antiaircraft guns. Barred by his non-combatant status from actively participating in keeping the guns firing, he gave his moral support to the ammunition bearers

✧✧✧

I can always make it a rule to get there first with the most men.

- Nathan Bedford Forrest

✧✧✧

What gets measured gets done; what gets rewarded gets done repeatedly.

- Barcy C. Fox

✧✧✧

There never was a good war, or a bad peace.

- Benjamin Franklin

✧✧✧

All wars are follies, very expensive and very mischievous ones. In my opinion, there never was a good war or a bad peace. When will mankind be convinced and agree to settle their difficulties by arbitration?

- Benjamin Franklin

✧✧✧

Use problems as opportunities to change our lives. Problems call forth our courage and wisdom.

- Benjamin Franklin

✧✧✧

Things that hurt, instruct.

- Benjamin Franklin

✧✧✧

And where is the Prince who can afford to so cover his country with troops for its defence, as that ten thousand men descending from the clouds, might not in many places do an infinite deal of mischief, before a force could be brought together to repel them?

- Benjamin Franklin, 1784

✧✧✧

For want of a nail, the shoe was lost; For want of a shoe, the horse was lost; For want of a horse, the rider was lost; For want of a rider, the battle was lost.

- Benjamin Franklin

✧✧✧

They that can give up essential liberty to purchase a little temporary safety, deserve neither liberty nor safety.

- Benjamin Franklin

✧✧✧

We must all hang together, or assuredly we shall all hang separately.

- Benjamin Franklin

✧✧✧

A defensive war is apt to betray us into too frequent detachment. Those generals who have had but little experience attempt to protect every point, while those who are better acquainted with their profession, having only the capital object in view, guard against a decisive blow, and acquiesce in small misfortunes to avoid greater.

- Frederick the Great

✧✧✧

The most certain way of insuring victory is to march briskly and in good order against the enemy, always endeavouring to gain ground.

- Frederick the Great

✧✧✧

A ruler is treated with no consideration if he does not have troops of his own. It is these, thank God! That have made me considerable since the time I began to have them.

- Frederick the Great

✧✧✧

Religion is the idol of the mob it adores everything it does not understand.

- Frederick the Great

✧✧✧

It is your attitude, and the suspicion that you are maturing the boldest designs against him, that imposes on your enemy.

- Frederick the Great

✧✧✧

Nothing is more difficult, almost impossible, than to prevent the passage of a river line, especially if the front is too extended. I should not care to be charged with such a mission.

- Frederick the Great

✧✧✧

The line is useless because it stretches over more ground than there are troops to defend it. If it is attacked in several places it can be over-whelmed with certainty. It does not, therefore, secure the area and merely deprives the defending troops of reputation and honour.

- Frederick the Great

✧✧✧

God is always with the strongest battalion.

- Frederick the Great

✧✧✧

Do not forget your dogs of war, your big guns, which are the most-to-be respected arguments of the rights of kings.

- Frederick the Great

✧✧✧

If my soldiers were to begin to think, not one would remain in the ranks.

- Frederick the Great

✧✧✧

All in all it seems, that an unusual spirit of independence of those above and acceptance of responsibility, has grown up throughout the Prussian officer corps as it has in no other army .... Prussian officers will not stand for being hemmed in by rules and stereotypes as happens in Russia, Austria or Britain ... we follow the more natural course of giving scope to every individual's talent, of using a looser rein. We back up every success as a matter of course—even if it runs counter to the intentions of the Commander-in-Chief ... the subordinate commander exploits every advantage by taking initiatives off his own bat, without his superior's knowledge or approval.

- Frederick the Great

✧✧✧

Everything which the enemy least expects will succeed the best.

- Frederick the Great, 1712–1786
Instructions to his Generals

✧✧✧

By push of bayonets, no firing until you see the whites in their eyes!

- Frederick the Great (1712–1786), at Prague, 1757

✧✧✧

Constitutions are made of paper; Bayonets are made of steel.

- French Aphorism

✧✧✧

The first human being who hurled an insult instead of a stone was the founder of civilization.

- Sigmund Freud

✧✧✧

Men are strong so long as they represent a strong idea; they become powerless when they oppose it.

- Sigmund Freud

✧✧✧

The direct use of force is such a poor solution to any problem, it is generally employed only by small children and large nations.

- David Friedman

✧✧✧

There are four ways in which you can spend money. You can spend your own money on yourself. When you do that, why then you really watch out what you're doing, and you try to get the most for your money. Then you can spend your own money on somebody else. For example, I buy a birthday present for someone. Well, then I'm not so careful about the content of the present, but I'm very careful about the cost. Then, I can spend somebody else's money on myself. And if I spend somebody else's money on myself, then I'm sure going to have a good lunch! Finally, I can spend somebody else's money on somebody else. And if I spend somebody else's money on somebody else, I'm not concerned about how much it is, and I'm not concerned about what I get. And that's government.

- Milton Friedman

✧✧✧

I will ignore all ideas for new works and engines of war, the invention of which has reached its limits and for whose improvement I see no further hope.

- Julius Frontinus

Chief military engineer to the Emperor Vespasian, c. AD 70.

✧✧✧

Two roads diverged in a wood and I—I took the one less travelled by, and that has made all the difference.

- Robert Frost

✧✧✧

The victor will be the one who summons the determination to attack; the side which only defends is doomed to inevitable defeat.

- MV Frunze

✧✧✧

There is no type of human endeavour where it is so important that the leader understands all phases of his job as that of the profession of arms.

- Major General James Fry

✧✧✧

What we may be witnessing is not just the end of the Cold War, or the passing of a particular period of postwar history, but the end of history as such... That is, the end point of mankind's ideological evolution and the universalization of Western liberal democracy as the final form of human government.

- Francis Fukuyama
The End of History and the Last Man

✧✧✧

There is, ... no single global strategy that works in terms of democratic openness. Sometimes it happens from the bottom up and sometimes it happens from the up down, and to be successful it usually has to work in both ways. There has to be elite that wants change, though that desire can be supported and driven by popular participation. For example in Chile, the Philippines and Korea it required pressure on leaders on top to open up their systems and those pressures couldn't have come only from civil society. In Ukraine and Georgia on the other hand there was obviously a big push from below—pressure in both directions is necessary. There is not one single strategy that produces democratic transition.

- Francis Fukuyama

✧✧✧

It's a really big mistake to think democratization is a good tool to fight terrorism."

- Francis Fukuyama

✧✧✧

The totality of U.S. military interventions have not left lasting, meaningful democratic institutions.

- Francis Fukuyama

✧✧✧

"War" is the wrong metaphor for the broader struggle, since wars are fought at full intensity and have clear beginnings and endings. Meeting the jihadist challenge is more of a "long, twilight struggle" whose core is not a military campaign but a political contest for the hearts and minds of ordinary Muslims around the world.

- Francis Fukuyama

✧✧✧

Air warfare is a shot through the brain, not a hacking to pieces of the enemy's body.

- Major General J.F.C. Fuller, British Army

✧✧✧

The object of all military training is to prepare the soldier for the next war, for it is the only possible war in which he can fight.

- Major General J.F.C. Fuller, British Army

✧✧✧

Artillery conquers and infantry occupies.

- Major General J.F.C. Fuller, British Army

✧✧✧

To me our bombing policy appears to be suicidal. Not because it does not do vast damage to our enemy, it does; but because, simultaneously, it does vast damage to our peace aim, unless that aim is mutual economic and social annihilation.

- Major General J.F.C. Fuller

✧✧✧

As the aeroplane is the most mobile weapon we possess, it is destined to become the dominant offensive arm of the future.

- Major General J.F.C. Fuller

✧✧✧

If you wish to understand war, examine the nature of surprise in its thousand and one forms as it pursues its restless course through history.

- Major General J.F.C. Fuller

✧✧✧

In the World War nothing was more dreadful to witness than a chain of men starting with a battalion commander and ending with an army commander sitting in telephone boxes, improvised or actual, talking, talking, talking, in place of leading, leading, leading.

- Major General J.F.C. Fuller

✧✧✧

It is absolutely true in war, were other things equal, that numbers, whether men, shells, bombs, etc., would be supreme. Yet it is also absolutely true that other things are never equal and can never be equal.

- Major General J.F.C. Fuller

✧✧✧

What is the soldier? ... Here we have the true answer to our question. It is not drill or uniform, badges or weapons, which make the soldier, but that spirit of self-sacrifice for a cause which he instinctively feels is a just one, which urges him on to a goal which he may never reach, or reaching it, may receive no further reward than the knowledge that through his efforts he has added to the greatness of the country and happiness and stability of his race.

- Colonel J.F.C. Fuller, D.S.O.
"Moral, Instruction and Leadership," Journal of the Royal United Services Institution, Vol. LXV, February to November, 1920

✧✧✧

What does discipline depend on? Will power and knowledge. You can do nothing without these two. A man who has no will power is an idiot; a man who has no knowledge is a fool. If a situation finds you with an empty skull it is too late to fill it. To go into action ignorant of what is required of you is as disastrous as to go into action without ammunition.

- Colonel J.F.C. Fuller, D.S.O.
"Moral, Instruction and Leadership," Journal of the Royal United Services Institution, Vol. LXV, February to November, 1920

✧✧✧

Intelligent obedience is but another name for initiative, and of all the qualities a soldier must possess, initiative will prove the most useful or the most dangerous according to its application. Initiative is really obedience without orders, that is, obedience with reference to the general plan and object of the operations as governed by the conditions of the moment.

- Colonel J.F.C. Fuller, D.S.O.
"Moral, Instruction and Leadership," Journal of the Royal United Services Institution, Vol. LXV, February to November, 1920

✧✧✧

A soldier may lose everything, including his life, but he may not lose his honour. Honour means loyalty not only to his King and Country, but to his officers and his comrades. It means truthfulness, thoughtfulness and manliness.

- Colonel J.F.C. Fuller, D.S.O.
"Moral, Instruction and Leadership," Journal of the Royal United Services Institution, Vol. LXV, February to November, 1920

✧✧✧

Grand strategy depends upon policy, and policy upon political knowledge of war. If this knowledge is nil, policy will be nil, and grand strategy will follow suit. In other words, generalship becomes impossible.

- J.F.C. Fuller (The Army in My Time)

✧✧✧

As the aeroplane is the most mobile weapon we possess, it is destined to become the dominant offensive arm of the future.

- J.F.C. Fuller (The Army in My Time)

✧✧✧

As the war of machines is less destructive to life and property, and more destructive to will and nerves, the tactical object of war will become demoralisation and disorganisation rather than destruction and annihilation.

- J.F.C. Fuller (The Army in My Time)

✧✧✧

The military object will not be to assault the outer casing of the enemy's armed forces, but the internal organs of command, intercommunication and supply. And as the whole of these and the fighting troops depend for their existence on the maintenance of political authority, and as political authority can be changed by the will of the people, the civil population becomes the main target.

- J.F.C. Fuller (The Army in My Time)

✧✧✧

To me our bombing policy appears to be suicidal. Not because it does not do vast damage to our enemy-it does; but because, simultaneously, it does vast damage to our peace aim, unless that aim is mutual economic and social annihilation.

- J.F.C. Fuller (Thunderbolts)

✧✧✧

We should speak to them [enlisted soldiers] as if they were our equals, but never as if we were theirs.

- *J.F.C. Fuller (How to Train an Army)*

✧✧✧

The more people as a whole are brought to realize the truth about the war-pleasant and unpleasant-the better are they place to accept its risks, endure its consequences, and turn to their advantage the facts which truth reveals.

- *J.F.C. Fuller (Thunderbolts)*

✧✧✧

It is absolutely true in war, were other things equal, that numbers-whether men, shells, bombs, etc.-would be supreme. Yet it is also absolutely true that other things are never equal and can never be equal.

- *J.F.C. Fuller (Thunderbolts)*

✧✧✧

The influence of the spirit of nationality, that is of democracy, on war is profound, as also were the influences of science and industrial development. The first emotionalized war and, consequently, brutalized it; and the second delivered into the hands of the masses more and more deadly means of destruction.

- *J.F.C. Fuller (War and Western Civilization)*

✧✧✧

National armies fight nations, royal armies fight their like, the first obey a mob-always demented, and the second a king-generally sane.

- *J.F.C. Fuller (War and Western Civilization)*

✧✧✧

Wars are not fought on drawing boards, nor are victories to be measured in pints. War is every whit as much a game of chance as a game of calculation: it is poker as well as chess, and even in chess there is no slide-rule certainty.

- *J.F.C. Fuller (Thunderbolts)*

✧✧✧

Whilst the nineteenth-century idea of warfare was founded on the principle of the concentration of force, because of air power the principle which will control the twentieth-century idea will be that of distribution of force, which is a sound strategical principle when the forces distributed possess so high a mobility that they can rapidly be concentrated at any desired point.

- *J.F.C. Fuller (The Army in My Time)*

✧✧✧

In the old strategy the leading idea was to concentrate superiority of force against a decisive point, in the new it is to distribute it in such a way that the enemy will be entangled in a tactical net. Again, whilst the former aimed at cutting an enemy off from his communications, the latter aims at cutting communications off from the enemy.

- *J.F.C. Fuller (The Army in My Time)*

✧✧✧

When, in 1898, I joined the Army, though a normally indifferently educated young Englishman, I was appalled by the ignorance which surrounded me and the immense military value attached to it.

- J.F.C. Fuller (The Army in My Time)

✧✧✧

Re. Britain's unpreparedness for the Boer War: The fact is we had made up our minds to play whist, and when we sat down we found that the game was poker.

- J.F.C. Fuller (The Army in My Time)

✧✧✧

Every pioneer is somewhat of a martyr, and every martyr somewhat of a firebrand who kills with ridicule as well as with reason.

- J.F.C. Fuller (Foundations of the Science of War)

✧✧✧

Out-of-date theories have consistently proved the ruin of armies.

- J.F.C. Fuller (War and Western Civilization)

✧✧✧

So it happened that the spirit of the unlimited offensive smote us like the Black Death: we began once again to think in terms of numbers in place of skill and of quantity in place of quality.

- J.F.C. Fuller (The Army in My Time)

✧✧✧

The principles of war ..... be reduced to three groups, namely, principles of control, resistance, and pressure, and finally to one law—the law of economy of force. Thus the system evolved from six principles in 1912 rose to eight in 1915, to, virtually, nineteen in 1923, and then descended to nine in 1925, with the added advantage that these nine can be merged into three, and these three into one law.

The Nine Principles have been expressed in various ways, but my arrangement is as follows:

1. Direction.
2. Offensive action.
3. Surprise.
4. Concentration.
5. Distribution.
6. Security.
7. Mobility.
8. Endurance.
9. Determination.

- Major General JFC Fuller (1925)

✧✧✧

Either war is obsolete or men are.

- R. Buckminster Fuller

✧✧✧

Many would be cowards if they had courage enough.

- Thomas Fuller, 1608-1681

✧✧✧

It is war that shapes peace, and armament that shapes war.

- Thomas Fuller

✧✧✧

G

The wingman is absolutely indispensable. I look after the wingman. The wingman looks after me. It's another set of eyes protecting you. That the defensive part. Offensively, it gives you a lot more firepower. We work together. We fight together. The wingman knows what his responsibilities are, and knows what mine are. Wars are not won by individuals. They're won by teams.

- Lt. Col. Francis S. "Gabby" Gabreski, US Air Force
28 victories in WWII and 6.5 MiGs over Korea.

✧✧✧

The air force's real strength lies in its people. The mission is not done by machines, it is done by people. The best weapons are of little value without trained and motivated people to operate and support them.

- General Charles Gabriel, US Air Force

✧✧✧

Integrity is the fundamental premise of service in a free society. Without integrity, the moral pillars of our military strength-public trust and self-respect are lost.

- General Charles A. Gabriel

✧✧✧

The Greeks developed the notion of military service as a moral obligation of the citizenship that rests at the base of nationalism. Without this, the modern nation state could not have emerged in the form that it did.

- Richard A. Gabriel, The Culture of War, New York) 1990

✧✧✧

During the Battle of Britain the question "fighter or fighter-bomber?" had been decided once and for all: The fighter can only be used as a bomb carrier with lasting effect when sufficient air superiority has been won.

- General Adolf Galland, Luftwaffe, 'The First and the Last,' 1954.

✧✧✧

Never abandon the possibility of attack. Attack even from a position of inferiority, to disrupt the enemy's plans. This often results in improving one's own position.

- General Adolf Galland, Luftwaffe.

✧✧✧

To use a fighter as a fighter-bomber when the strength of the fighter arm is inadequate to achieve air superiority is putting the cart before the horse.

- General Adolf Galland, Luftwaffe

✧✧✧

Superior technical achievements—used correctly both strategically and tactically—can beat any quantity numerically many times stronger yet technically inferior.

- General Adolf Galland, Luftwaffe.

✧✧✧

If we should have to fight, we should be prepared to do so from the neck up instead of from the neck down.

- General Adolf Galland, Luftwaffe.

✧✧✧

The study of money, above all other fields in economics, is one in which complexity is used to disguise truth or to evade truth, not to reveal it. The process by which banks create money is so simple the mind is repelled. With something so important, a deeper mystery seems only decent.

- John Kenneth Galbraith (1908–2006), Economist

✧✧✧

Under capitalism, man exploits man. Under communism, it's just the opposite.

- John Kenneth Galbraith

✧✧✧

The modern conservative is engaged in one of man's oldest exercises in moral philosophy; that is, the search for a superior moral justification for selfishness.

- John Kenneth Galbraith

✧✧✧

All of the great leaders have had one characteristic in common: it was the willingness to confront unequivocally the major anxiety of their people in their time. This, and not much else, is the essence of leadership.

- John Kenneth Galbraith

✧✧✧

If all else fails, immortality can always be assured by spectacular error.

- John Kenneth Galbraith

✧✧✧

If wrinkles must be written upon our brows, let them not be written upon the heart. The spirit should never grow old.

- John Kenneth Galbraith

✧✧✧

In all life one should comfort the afflicted, but verily, also, one should afflict the comfortable, and especially when they are comfortably, contentedly, even happily wrong.

- John Kenneth Galbraith

✧✧✧

In any great organization it is far, far safer to be wrong with the majority than to be right alone.

- John Kenneth Galbraith

✧✧✧

Meetings are a great trap. Soon you find yourself trying to get agreement and then the people who disagree come to think they have a right to be persuaded. However, they are indispensable when you don't want to do anything.

- John Kenneth Galbraith

✧✧✧

Politics is not the art of the possible. It consists in choosing between the disastrous and the unpalatable.

- John Kenneth Galbraith

✧✧✧

One of the greatest pieces of economic wisdom is to know what you do not know.

- John Kenneth Galbraith

✧✧✧

The enemy of the conventional wisdom is not ideas but the march of events.

- John Kenneth Galbraith

✧✧✧

Condign power wins submission by the ability to impose an alternative to the preferences of the individual or group that is sufficiently unpleasant or painful so that these preferences are abandoned. There is an overtone of punishment. The expected rebuke is usually too harsh, so the individual will endure, submit, or give into the power from fear or threat. The individual is aware of the submission via compulsion.

- John Kenneth Galbraith
The Anatomy of Power

✧✧✧

Compensatory power wins submission by the offer of affirmative reward—by the giving of something of value to the individual so submitting. Payments, share, praise, money for services. The individual is aware of the submission for a reward.

- John Kenneth Galbraith
The Anatomy of Power

✧✧✧

Conditioned power wins submission by changing beliefs. Persuasion, education, habituation, social commitment to what seems natural, proper, right causes the individual to submit to the will of another or others. Submission reflects the preferred course; the fact of submission is not recognized. Conditioned power is central to the functioning of the modern economy and polity, and in capitalist and socialist countries alike.

- John Kenneth Galbraith
The Anatomy of Power

✧✧✧

The troops who fight the best do so, not because of their nationality, but because at that specific time they are the best trained, the best disciplined and the best led on the field.

- Strome Galloway, The General Who Never Was, 1981

✧✧✧

A nation' s strength ultimately consists in what it can do on its own, and not in what it can borrow from others.

- Indira Gandhi

✧✧✧

You cannot shake hands with a clenched fist.

- Indira Gandhi

✧✧✧

I awoke around 5 that morning after a fitful sleep to find my bag and I were soaked through. After mentally weeping for an hour or so, I perceived the rain was less persistent. "Aha", I thought, "if it stops I could get up and stamp around in the wind and start getting things dry". As the rain faded away, I began to gather together my shattered spirits. The rain stopped. I waited 10 minutes, just enjoying not getting any wetter. The rain started again. I pulled the sodden wretched bag over my head and smoked a cigarette. It was the last barrier between me and desperation.

- Captain I.R. Gardiner
A Personal Account of Operations on the Falklands Islands,
45 Commando Royal Marine, April-June 1982

✧✧✧

When waves of change appear, you can duck under the wave, stand fast against the wave, or, better yet, surf the wave.

- Bill Gates, founder of Microsoft

✧✧✧

As we look ahead into the next century, leaders will be those who empower others.

- Bill Gates

✧✧✧

If you can't make it good, at least make it look good.

- Bill Gates

✧✧✧

It's fine to celebrate success but it is more important to heed the lessons of failure.

- Bill Gates

✧✧✧

We always overestimate the change that will occur in the next two years and underestimate the change that will occur in the next ten. Don't let yourself be lulled into inaction.

- Bill Gates

✧✧✧

Having to do it for the first time in combat is a chastening experience, it gives a man religion.

- General J.M. Gavin, after his first combat paradrop, 1944

✧✧✧

The attacker who neglects vertical envelopment is outmoded;

The defender who digs in on a line is outflanked.

- General J.M. Gavin

✧✧✧

The battle of Arbela did not cover much more than a dozen square miles. The Battle of Zama (202 B.C.) in which the Roman Scipio decisively defeated the Carthaginian Hannibal and thus gave mastery of the Mediterranean to the Romans, extended over but a few square miles. It was one of the decisive battles of history as it led to Roman rule of the known world. The Battle of Normandy, the decisive battle for a lodgement on the continent of Europe, fought thousands of years later, extended over 10,000 square miles.

- General J.M. Gavin (War and Peace in the Space Age)

✧✧✧

A man's greatest moment in life is when his enemy lays vanquished, his village aflame, his herds driven before you and his weeping wives and daughters are clasped to your breast.

- Genghis Khan

✧✧✧

It is not sufficient that I succeed—all others must fail.

- Genghis Khan

✧✧✧

I am the punishment of God...If you had not committed great sins, God would not have sent a punishment like me upon you.

- Genghis Khan

✧✧✧

Anything can be achieved in small, deliberate steps. But there are times you need the courage to take a great leap; you can't cross a chasm in two small jumps.

- David Lloyd George

✧✧✧

You are not going to get peace with millions of armed men. The chariot of peace cannot advance over a road littered with cannon.

- David Lloyd George

✧✧✧

Believe me, Germany is unable to wage war.

- Former British Prime Minister David Lloyd George, 1 August 1934.

✧✧✧

We will squeeze the oranges until the pips squeak.

- David Lloyd George, British Prime Minister

✧✧✧

Diplomats were invented simply to waste time.

- David Lloyd George, British Prime Minister

❖❖❖

Wars teach us not to love our enemies, but to hate our allies.

- W. L. George

❖❖❖

History is accelerating. The pace is alarming. The direction is not entirely known. At this time of stress, the hard fact is that no power or combination of powers is prepared to take on its shoulders the responsibility of collective security worldwide.

- Boutros Boutrous Ghali, UN Secretary General
An Agenda for Peace

❖❖❖

The concept of peace is easy to grasp; that of international security is more complex, for a pattern of contradictions has arisen here as well. As major nuclear Powers have begun to negotiate arms reduction agreements, the proliferation of weapons of mass destruction threatens to increase and conventional arms continue to be amassed in many parts of the world. As racism becomes recognized for the destructive force it is and as apartheid is being dismantled, new racial tensions are rising and finding expression in violence. Technological advances are altering the nature and the expectation of life all over the globe. The revolution in communications has united the world in awareness, in aspiration and in greater solidarity against injustice. But progress also brings new risks for stability: ecological damage, disruption of family and community life, greater intrusion into the lives and rights of individuals.

- Boutros Boutros Ghali, UN Secretary General
An Agenda for Peace: Preventive diplomacy, peacemaking and peace-keeping (1992)

❖❖❖

We have entered a time of global transition marked by uniquely contradictory trends. Regional and continental associations of States are evolving ways to deepen cooperation and ease some of the contentious characteristics of sovereign and nationalistic rivalries. National boundaries are blurred by advanced communications and global commerce, and by the decisions of States to yield some sovereign prerogatives to larger, common political associations. At the same time, however, fierce new assertions of nationalism and sovereignty spring up, and the cohesion of States is threatened by brutal ethnic, religious, social, cultural or linguistic strife. Social peace is challenged on the one hand by new assertions of discrimination and exclusion and, on the other, by acts of terrorism seeking to undermine evolution and change through democratic means.

- Boutros Boutros Ghali, UN Secretary General
An Agenda for Peace: Preventive diplomacy, peacemaking and peace-keeping (1992)

❖❖❖

The enemy will pass slowly from the offensive to the defensive. The blitzkrieg will transform itself into a war of duration. Thus, the enemy will be caught in a dilemma: He has to drag out the war in order to win it, and does not possess, on the other hand, the psychological and political means to fight a long, drawn-out war.

- Vo Nguyen Giap
Vietnamese commanding general from 1944-1978,
on guerillas fighting a conventional Western army

✧✧✧

The nature and the very aim of the campaign the (colonialist) enemy is conducting oblige the enemy to split up his forces so as to be able to occupy the invaded territory...the enemy was thus faced with a contradiction: It was impossible for him to occupy the invaded territory without dividing his forces. By their dispersal, he created difficulties for himself. His scattered units thus became an easy prey for our troops and mobile forces.

- Vo Nguyen Giap

✧✧✧

It is easier to lead men to combat, stirring up their passion, than to restrain them and direct them toward the patient labours of peace.

- Andre Gide (1869–1951)
French author, winner of the Nobel Prize in literature in 1947

✧✧✧

Fish die belly-upward and rise to the surface; it is their way of falling.

- Andre Gide

✧✧✧

One does not discover new lands without consenting to lose sight of the shore for a very long time.

- Andre Gide

✧✧✧

It is better to be hated for what you are than to be loved for something you are not.

- Andre Gide

✧✧✧

Believe those who are seeking the truth. Doubt those who find it.

- Andre Gide

✧✧✧

A heart to resolve, a head to contrive, and a hand to execute.

- Edward Gibbon (1737–1794) English historian

✧✧✧

Every man who rises above the common level has received two educations: the first from his teachers; the second, more personal and important, from himself.

- Edward Gibbon

✧✧✧

History is little more than the register of the crimes, follies, and misfortunes of mankind.

- Edward Gibbon

✧✧✧

The courage of a soldier is found to be the cheapest and most common quality of human nature.

- Edward Gibbon

✧✧✧

The winds and the waves are always on the side of the ablest navigators.

- Edward Gibbon

✧✧✧

So long as mankind shall continue to lavish more praise upon its destroyers than upon its benefactors war shall remain the chief pursuit of ambitious minds.

- Edward Gibbon

✧✧✧

I never make the mistake of arguing with people for whose opinions I have no respect.

- Edward Gibbon

✧✧✧

[On ancient Athens]: In the end, more than freedom, they wanted security. They wanted a comfortable life, and they lost it all—security, comfort, and freedom. When the Athenians finally wanted not to give to society but for society to give to them, when the freedom they wished for most was freedom from responsibility, then Athens ceased to be free and was never free again.

- Edward Gibbon,
The Decline and Fall of the Roman Empire

✧✧✧

TOMMY ATKINS: The popular generic name for the British private soldier. In its origin the name dates from August 1815. (Waterloo year), when the War Office issued the first "Soldier's Account Book", which every soldier was provided with. The specimen form sent out with the book to show how details should be filled in, bore at the place where a man's signature was required the hypothetical name "Thomas Atkins", (or, alternatively, for illiterate men "Thomas Atkins X his mark") "Thomas Atkins" continued to appear in later editions of the Soldier's Account Book until comparatively recent times. It has now disappeared. A more or less current Service slang name from about 1830, the general popularity of the name "Tommy" for a soldier dates from about fifty years ago. Mr. Kipling's verses finally familiarized it all over the English-speaking world.

- Edward Fraser and John Gibbons
Soldier and Sailor Words and Phrases, 1925

✧✧✧

Take her down!

- Commander Howard Walter Gilmore
desperately wounded and unable to climb back into his submarine, USS Growler, in the face of an approaching Japanese gunboat 7 February 1943

✧✧✧

Here is my first principle of foreign policy: good government at home.

- William E. Gladstone (1809–1898) Four times Prime Minister of England

✧✧✧

Justice delayed is justice denied.

- William E. Gladstone

✧✧✧

If you are cold, tea will warm you; if you are too heated, it will cool you; if you are depressed, it will cheer you; if you are excited, it will calm you.

- William E. Gladstone

✧✧✧

No man ever became great or good except through many and great mistakes.

- William E. Gladstone

✧✧✧

Here is my first principle of foreign policy: good government at home.

- William E. Gladstone

✧✧✧

It is the duty of government to make it difficult for people to do wrong, easy to do right.

- William E. Gladstone

✧✧✧

Do you want total war? If necessary, do you want a war more total and radical than anything that we can even imagine today?

- Josef Goebbels,
Nazi Minister of Popular Enlightenment and Propaganda

✧✧✧

For the ambitious, initiative consists in seizing every opportunity to increase notoriety.

For disciplinarians, initiative on the part of subordinates is a misconception of their duties.

For imaginative people, initiative is the right to do anything which suddenly strikes them.

For lazy people initiative is the right to pass all irksome duty on to their subordinates.

For the easy-going, initiative consists in modifying to their liking any order they may receive.

For the timid, initiative is the right to shirk responsibility.

- Colonel F. Gory, "L'Initiative des Militaires," 1909

✧✧✧

He who stays on the defensive does not make war, he endures it.

- Field Marshal Colmar Baron von der Goltz, 1883

✧✧✧

Good leadership requires you to surround yourself with people of diverse perspective, who can disagree with you without fear of retaliation.

- Doris Kearns Goodwin

✧✧✧

The battle we are now approaching demands a colossal measure of production capacity. No limit on rearmament can be visualized. The only alternatives are victory or destruction... We live in a time when the final battle is in sight. We are ready on the threshold of mobilization and we are already at war. All that is lacking is the actual shooting.

- Reich Marshal Hermann Göering,
Commander-in-Chief of the Luftwaffe (1936)

✧✧✧

I believe this plan [raiding RAF airfields] would have been very successful, but as a result of the Fuhrer's speech about retribution, in which he asked that London be attacked immediately, I had to follow the other course. I wanted to attack the airfields first, thus creating a prerequisite for attacking London . . . I spoke with the Fuhrer about my plans in order to try to have him agree I should attack the first ring of RAF airfields around London, but he insisted he wanted to have London itself attacked for political reasons, and also for retribution.

I considered the attacks on London useless, and I told the Fuhrer again and again that inasmuch as I knew the English people as well as I did my own people, I could never force them to their knees by attacking London. We might be able to subdue the Dutch people by such measures but not the British.

- Reichmarschall Hermann Goering,
International Military Tribunal Nuremberg, 1946.

✧✧✧

Above all, I shall see to it that the enemy will not be able to drop any bombs.

- Hermann Goering, German Air Force Minister
German original: "Vor allem werde ich dafur sorgen,
dass der Feind keint Bomben werfen Kann."

✧✧✧

No enemy bomber can reach the Ruhr. If one reaches the Ruhr, my name is not Goering. You may call me Meyer.

- Herman Goering, German Air Force Minister,
addressing the German Air force, September 1939.

✧✧✧

The true master is the one who knows the art of restraint.

- William Goethe

✧✧✧

There is nothing more fearful than ignorance in action.

- Johann Wolfgang von Goethe
Proverbs in Prose

✧✧✧

Are you in earnest; seize this very minute.
What you can do or dream you can, begin it;
Boldness has genius power and magic in it;
Only engage, and the mind grows heated;
Begin, and then the work will be completed.

- Johann Wolfgang von Goethe

✧✧✧

Those who have not yet realized danger are generally the bravest soldiers.

- Colmar von der Goltz, German Field Marshal, World War I

✧✧✧

The most difficult thing about planning against the Americans, is that they do not read their own doctrine, and they would feel no particular obligation to follow it if they did.

- Admiral Sergei I. Gorshkov (1910–1988)
father of the Russian blue water navy

✧✧✧

A subaltern [in the 1920s and 30s] could not marry without obtaining permission from his commanding officer. The reasoning was that he might get into debt, for he would not be entitled to either married quarters or a marriage allowance. And he would spend too much time with his wife and neglect his horses, guns, and men. The mess, too, would suffer from his absence. Subalterns were at the bottom of the regimental totem pole. In his senior term at the RMC he had been warned that the first claim on the gunner officer's loyalty and attention was the infantry or cavalry unit that he was assigned to support. Next came the guns with which he gave support, the horses that took the guns into action, and the men in the gun detachments. The battery and regiment embraced all these, and the private affairs or preferences of young subalterns were of no importance.

- Dominick Graham
The Price of Command; A Biography of General Guy Simonds

✧✧✧

I have never advocated war except as a means of peace.

- Ulysses S. Grant (1822–1885)
18th President of the United States (1869–1877)

✧✧✧

In every battle there comes a time when both sides consider themselves beaten, then he who continues the attack wins.

- Ulysses S. Grant

✧✧✧

If men make war in slavish obedience to rules, they will fail.

- Ulysses S. Grant

✧✧✧

The art of war is simple enough. Find out where your enemy is. Get at him as soon as you can. Strike him as hard as you can, and keep moving on.

- Ulysses S. Grant

✧✧✧

My failures have been errors in judgment, not of intent.

- Ulysses S. Grant

✧✧✧

The right of revolution is an inherent one. When people are oppressed by their government, it is a natural right they enjoy to relieve themselves of oppression, if they are strong enough, whether by withdrawal from it, or by overthrowing it and substituting a government more acceptable.

- Ulysses S. Grant

✧✧✧

There never was a time when, in my opinion, some way could not be found to prevent the drawing of the sword.

- Ulysses S. Grant

✧✧✧

The art of war is simple enough. Find out where your enemy is. Get at him as soon as you can. Strike him as hard as you can, and keep moving.

- Ulysses S. Grant

✧✧✧

I don't run democracy. I train troops to defend democracy and I happen to be their surrogate father and mother as well as their commanding general.

- Lt Gen Alfred M. Gray, USMC

✧✧✧

[I go] where the sound of thunder is.

- Lt Gen Alfred M. Gray

✧✧✧

There's no such thing as a crowded battlefield. Battlefields are lonely places.

- Lt Gen Alfred M. Gray

✧✧✧

High-tech transformation will have only modest value, because war is a duel and all of America's foes out to 2020 will be significantly asymmetrical. The more intelligent among them, as well as the geographically more fortunate and the luckier, will pursue ways of war that do not test US strengths. Second, the military potential of this transformation, as with all past transformations, is being undercut by the unstoppable processes of diffusion which spread technology and ideas. Third, the transformation that is being sought appears to be oblivious to the fact claimed here already, that there is more to war than warfare. War is about the peace it will shape.

- Colin S. Gray

✧✧✧

The time will come, when thou shalt lift thine eyes
To watch a long-drawn battle in the skies.
While aged peasants, too amazed for words,
Stare at the flying fleets of wondrous birds.

England, so long mistress of the sea,
Where winds and waves confess her sovereignty,
Her ancient triumphs yet on high shall bear
And reign the sovereign of the conquered air.

- Thomas Gray, 1737

✧✧✧

Adventure is not outside a man it is within.

- David Grayson

✧✧✧

A militarist is one whose outlook covers only purely military matters, taking into no account, and being incapable of taking into account, the soul which exists in nations.

Sir Edward Grey, British Foreign Minister, 1914

✧✧✧

The mind of the enemy and the will of his leaders is a target of far more importance than the bodies of his troops.

- Brigadier General S.B. Griffith, II, US Marine Corps

✧✧✧

There is a common law among nations, which is valid alike for war and in war, I have had many and weighty reasons for undertaking to write upon the subject. Throughout the Christian world I observed a lack of restraint in relation to war, such as even barbarous races should be ashamed of; I observed that men rush to arms for slight causes, or no cause at all, and that when arms have once been taken up there is no longer any respect for law, divine or human; it is as if, in accordance with a general decree, frenzy had openly been let loose for the committing of all crimes.

- Hugo Grotius (1583–1645)

✧✧✧

A man cannot govern a nation if he cannot govern a city; he cannot govern a city if he cannot govern a family; he cannot govern a family unless he can govern himself; and he cannot govern himself unless his passions are subject to reason.

- Hugo Grotius

✧✧✧

Even God cannot make two times two not make four.

- Hugo Grotius

✧✧✧

Not to know certain things is a great part of wisdom.

- Hugo Grotius

✧✧✧

Liberty is the power that we have over ourselves.

- Hugo Grotius

✧✧✧

Any logical model of reality is incomplete (and possibly inconsistent) and must be continuously refined/adapted in the face of new observations.

- Gödel's Incompleteness Theorem

✧✧✧

Boldness is good in all things and nothing is well done which is fearfully done.

- Stefano Guazzo

✧✧✧

A unit that wants to win battles must excel in its ability to move, shoot and communicate.

- Field Marshal Heinz Guderian (1888–1954)

✧✧✧

When the situation is obscure, attack!

- Field Marshal Heinz Guderian (1888-1954)

✧✧✧

Our schools should train leaders and commanders in how to train for battle, not just for battle.

- Field Marshal Heinz Guderian

✧✧✧

Start each period of instruction with two statements: Tell them what they are going to learn and why they need to know it.

- Field Marshal Heinz Guderian

✧✧✧

We have severely underestimated the Russians, the extent of the country and the treachery of the climate. This is the revenge of reality.

- Field Marshal Heinz Guderian, letter to his wife 1941

✧✧✧

Logistics is the ball and chain of armoured warfare.

- Field Marshal Heinz Guderian

✧✧✧

There are no desperate situations, there are only desperate people.

- Field Marshal Heinz Guderian

✧✧✧

I had complete trust in their competence and reliability. They knew my views and shared my belief .... and could follow it even though they may receive no orders for long periods of time once the attack was launched.

- Field Marshal Heinz Guderian

✧✧✧

If the tanks succeed, then victory follows.

- Field Marshal Heinz Guderian

✧✧✧

In this year (1929) I became convinced that tanks working on their own or in conjunction with infantry could never achieve decisive importance. My historical studies; the exercises carried out in England and our own experience with mock-ups had persuaded me that the tanks would never be able to produce their full effect until weapons on whose support they must inevitably rely were brought up to their standard of speed and of cross-country performance. In such formation of all arms, the tanks must play primary role, the other weapons being subordinated to the requirements of the armor. It would be wrong to include tanks in infantry divisions: what was needed were armored divisions which would include all the supporting arms needed to fight with full effect.

- Field Marshal Heinz Guderian, Panzer Leader

✧✧✧

You hit somebody with your fist and not with your fingers spread.

- Field Marshal Heinz Guderian

✧✧✧

Courage is a special kind of knowledge: the knowledge of how to fear what ought to be feared and how not to fear what ought not to be feared.

- David Ben Gurion (1886-1973), First Prime Minister of Israel

✧✧✧

If an expert says it can't be done, get another expert.

- David Ben Gurion

✧✧✧

...am going on a raid this afternoon... there is a possibility I won't return.. do not worry about me as everyone has to leave this Earth one way or another, and this is the way I have selected. If after this terrible war is over, the world emerges a saner place...pogroms and persecutions halted, then, I'm glad I gave my efforts with thousands of others for such a cause.

- Sergeant Carl Goldman
U.S. Army Air Forces, WWII, B-17 Gunner, Killed in Action over Western Europe...from a letter to his parents

✧✧✧

Owing to the development of aviation war has altered in character. Hitherto primarily an affair of "fronts," it will henceforth be primarily an affair of "areas".

- Brig Gen P.R.C. Groves, RAF, 1922

✧✧✧

None would dare to aver that there will be no more war, for if that were so then the problem would have been forever solved; and if wars there are to be they will be lost or won in the air.

- Brig Gen P.R.C. Groves, RAF, 1922

✧✧✧

In the long run the only safe and reliable foundation of national air power lies in commercial air development.

- P.R.C. Groves (Behind the Smoke Screen)

✧✧✧

A large system of commercial aviation will remain an abiding support for the national aircraft industry and will at the same time provide those reserves of personnel which are essential for the expansion of aerial strength in case of need.

- P.R.C. Groves (Behind the Smoke Screen)

✧✧✧

A country which cannot defend itself from aerial attack will find its air bases, its munitions centres, its military depots, its shipyards, and its great cities subjected to a devastating rain of bombs within a few hours of the declaration of hostilities.

- Brig Gen P.R.C. Groves, RAF, 1922

✧✧✧

So long as we allow tradition, in place of science, to dictate the allocation of our defensive expenditure, we shall dissipate vast sums on pseudo-security, enfeeble our diplomacy and betray the cause of peace.

- P.R.C. Groves (Behind the Smoke Screen)

✧✧✧

One of the most obvious lessons of history is that the preservation of States and civilization lies in the ability to allow for changed conditions.

- P.R.C. Groves (Behind the Smoke Screen)

✧✧✧

H

The power of example is very important to people under stress.

- General Sir John Hackett, British Army (1910–1997)

✧✧✧

Bad people can be good doctors or lawyers, but what the bad man cannot be is a good sailor, or soldier or airman.

- General Sir John Hackett, British Army

✧✧✧

The essential basis of military life is the ordered application of force under an unlimited liability. It is the unlimited liability which sets the man who embraces this life somewhat apart. He will be (or should be) always a citizen. So long as he serves he will never be a civilian.

- General Sir John Hackett, British Army

✧✧✧

Bravery is being the only one who knows you're afraid.

- Col. David H. Hackworth

✧✧✧

Bombardment from the air is legitimate only when directed at a military objective, the destruction or injury of which would constitute a distinct military disadvantage to the belligerent.

- The Hague Convention of Jurists, 1923.

✧✧✧

The problem with sports and war is that God is on everyone's side.

- Duane Alan Hahn

✧✧✧

The warning message we sent the Russians was a calculated ambiguity that would be clearly understood.

- General Alexander Haig (1924–2010)
US Secretary of State

✧✧✧

We didn't lose Vietnam. We quit Vietnam.

- General Alexander Haig

✧✧✧

That's not a lie, it's a terminological inexactitude. Also, a tactical misrepresentation.

- General Alexander Haig

✧✧✧

Israel is the largest American aircraft carrier in the world that cannot be sunk, does not carry even one American soldier, and is located in a critical region for American national security.

- General Alexander Haig

✧✧✧

The fundamental task of diplomacy is to strip policy of its ambiguity.

- General Alexander Haig

✧✧✧

I only regret that I have but one life to lose for my country.

- Nathan Hale

✧✧✧

There can be no peace but that which is forced by the sword.

- Henry Halleck
19th century American military writer and Civil War general

✧✧✧

I never trust a fighting man who doesn't smoke or drink.

- Admiral William Halsey (1882–1959)

✧✧✧

The four machines that won the War in the Pacific were the submarine, radar, the airplane and the bulldozer.

- Admiral William Halsey

✧✧✧

There are no great people in this world, only great challenges which ordinary people rise to meet.

- Admiral William Halsey

✧✧✧

Peacekeeping is not a job for soldiers, but only soldiers can do it.

- Dag Hammarskjöld, Former UN Secretary-General

✧✧✧

The pursuit of peace and progress cannot end in a few years in either victory or defeat. The pursuit of peace and progress, with its trials and its errors, its successes and its setbacks, can never be relaxed and never abandoned.

- Dag Hammarskjöld, Former UN Secretary-General

✧✧✧

Everything will be all right—you know when? When people, just people, stop thinking of the United Nations as a weird Picasso abstraction and see it as a drawing they made themselves.

- Dag Hammarskjöld, Former UN Secretary-General

✧✧✧

I divide officers into four classes—the clever, the lazy, the stupid and the industrious. Each officer possesses at least two of these qualities. Those who are clever and industrious are fitted for the high staff appointments. Use can be made of those who are stupid and lazy. The man who is clever and lazy is fit for the very highest commands. He has the temperament and the requisite nerves to deal with all situations. But whoever is stupid and industrious must be removed immediately.

- General Baron Kurt von Hammerstein (1878-1943);
German Chief of Army Command (1930-33)

✧✧✧

So it comes that the balance point of discipline has, at least in the British Army, insensibly shifted during the past twenty years. Soldiers of all ranks are being taught the reason why orders must be obeyed; why an unwise command wisely and unanimously carried out will carry further than a wise order hesitatingly executed; why individuals must stoop if they wish the nation to conquer. Force of habit of mind has replaced force of habit of body.

- General Ian Hamilton,
The Soul and Body of an Army, 1921

✧✧✧

Democracy has no convictions for which people would be willing to stake their lives.

- Dr. Ernst Hanfstaengl

✧✧✧

Aut Viam Inveniam Aut Faciam

We will either find a way or make one.

- Hannibal (247-183 BC), Carthaginian general
(when told it was impossible to cross the Alps by elephant)

✧✧✧

God has given to man no sharper spur to victory than contempt of death.

- Hannibal

✧✧✧

War will cease when men refuse to fight.

- Fridtjof Hansen

✧✧✧

Violence is the last refuge of the incompetent.

- Salvor Hardin

✧✧✧

We are going to scourge the Third Reich from end to end. We are bombing Germany city by city and ever more terribly in order to make it impossible for her to go in with the war. That is our object, and we shall pursue it relentlessly.

- Marshal of the Royal Air Force Arthur 'Bomber' Harris, July 1942

✧✧✧

Victory, speedy and complete, awaits the side which first employs air power as it should be employed. Germany, entangled in the meshes of vast land campaigns, cannot now disengage her air power for a strategically proper application. She missed victory through air power by a hair's breadth in 1940. . . . We ourselves are now at the crossroads.

- Air Marshal Sir Arthur "Bomber" Harris,
opening of letter to Winston Churchill, 17 June 1942

✧✧✧

There are a lot of people who say that bombing cannot win the war. My reply to that is that it has never been tried. . . and we shall see.

- Marshal of the Royal Air Force Sir Arthur "Bomber" Harris

✧✧✧

War is a nasty, dirty, rotten business. It's all right for the Navy to blockade a city, to starve the inhabitants to death. But there is something wrong, not nice, about bombing that city.

- Marshal of the Royal Air Force Arthur 'Bomber' Harris

✧✧✧

No tradition is worth having in a fighting force except a tradition of success.

- Marshal of the Royal Air Force Sir Arthur Harris

✧✧✧

A rifleman's observations on his officers

It is, indeed, singular, how a man loses or gains caste with his comrades from his behaviour, and how closely he is observed in the field. The officers, too, are commented upon and closely observed. The men are very proud of those who are brave in the field, and kind and considerate to the soldiers under them. An act of kindness done by an officer has often during the battle been the cause of his life being saved. Nay, whatever folks may say upon the matter, I know from experience, that in our army the men like best to be officered by gentlemen, men whose education has rendered them more kind in manners than your coarse officer, sprung from obscure origin, and whose style is brutal and overbearing.

- Recollections of Rifleman Harris
Edited by Captain Henry Curling, London 1848

✧✧✧

I have always regarded the forward edge of battle as the most exclusive club in the world.

- Sir Brian Harrock

✧✧✧

To adopt the method of attrition is not only a confession of stupidity, but a waste of strength, endangering both the chances during the combat and the profit of victory.

- B.H. Liddell Hart (Thoughts on War)

✧✧✧

In the last war, air-power forfeited much of its effect from being kept in separate packets like the parts of an army, with a consequent dispersion of effort and frittering of effect.

- B.H. Liddell Hart (Thoughts on War)

✧✧✧

The history of mankind is the history of thought-of the gradual ascendancy of mind over matter: the subjugation of brute force by intelligence.

- B.H. Liddell Hart (Thoughts on War)

✧✧✧

Air power is, above all, a psychological weapon-and only short-sighted soldiers, too battle-minded, underrate the predominance of psychological factors in war.

- B.H. Liddell Hart (Thoughts on War)

✧✧✧

Until we understand war in the fullest sense, which involves an understanding of men in war, among other elements, it seems to me that we have no more prospect of preventing war than the savage has of preventing plague.

- B.H. Liddell Hart (Thoughts on War)

✧✧✧

There is no excuse for any literate person if he is less than three thousand years old in mind.

- B.H. Liddell Hart (Thoughts on War)

✧✧✧

Of all qualities in war it is speed which is dominant, speed both of mind and movement-without which hitting-power is valueless and with which it is multiplied.

- B.H. Liddell Hart (Thoughts on War)

✧✧✧

The profoundest truth of war is that the issue of battle is usually decided in the minds of the opposing commanders, not in the bodies of their men.

- B. H. Liddell Hart

✧✧✧

Every gain in speed increases not only the attacker's security but the defender's insecurity. For the higher the speed the greater the chance of, and scope for, surprise. Speed and surprise are not merely related; they are twins.

- B.H. Liddell Hart (Thoughts on War)

✧✧✧

Self-interest as well as humane reasons demand that warring nations should endeavour to gain their end-the moral subjugation of the enemy-with the infliction of the least possible permanent injury to life and industry. For the enemy of to-day is the customer of tomorrow, and often the ally of the future.

- B.H. Liddell Hart (Thoughts on War)

✧✧✧

Victory is not an end in itself. It is worse than useless if the end of the war finds you so exhausted that you are defeated in the peace. Wise statesmanship must aim at conserving strength so as to be still strong when peace is settled.

- B.H. Liddell Hart (Thoughts on War)

✧✧✧

War is "too serious a business" for the destinies of nations to be controlled by mere strategists. There is a need for the wider horizon of grand strategy, which embraces the state of peace that lies beyond every war.

- B.H. Liddell Hart (Thoughts on War)

✧✧✧

In peace we concentrate so much on tactics that we are apt to forget that it is merely the handmaiden of strategy.

- B.H. Liddell Hart (Thoughts on War)

✧✧✧

The dead hand of Clausewitz on the strategy of his country's *opponents* may well be counted as his supreme patriotic legacy.

- B.H. Liddell Hart (Thoughts on War)

✧✧✧

An important difference between a military operation and a surgical operation is that the patient is not tied down. But it is a common fault of generalship to assume that he is.

- B.H. Liddell Hart (Thoughts on War)

✧✧✧

Surprise is the master-key of war.

- B.H. Liddell Hart (Thoughts on War)

✧✧✧

Movement generates surprise, and surprise gives impetus to movement. For a movement which is accelerated or changes its direction inevitably carries with it a degree of surprise, even though it be unconcealed; while surprise smoothes the path of movement by hindering the enemy's counter-measures and counter-movements.

- B.H. Liddell Hart (How to Win Wars)

✧✧✧

The real target in war is the mind of the enemy commander, not the bodies of his troops. If we operate against his troops it is fundamentally for the effect that action will produce on the mind and will of the commander; indeed, the trend of warfare and the development of new weapons-aircraft and tanks-promise to give us increased and more direct opportunities of striking at this psychological target.

- B.H. Liddell Hart (Thoughts on War)

✧✧✧

In determining the role of mechanized forces, let us hark back to Napoleon's "the whole secret of the art of war lies in making oneself master of the communications.

- B.H. Liddell Hart (Thoughts on War)

✧✧✧

It is the function of grand strategy to discover and exploit the Achilles' heel of the enemy nation; to strike not against its strongest bulwark but against its most vulnerable spot.

- B.H. Liddell Hart (Thoughts on War)

✧✧✧

History shows us that no entirely new weapon has radically affected the course of any war; that the decisive weapon in a war has always been known, if but in a crude and undeveloped form, in the previous war.

- B.H. Liddell Hart (Thoughts on War)

✧✧✧

For Clausewitz, in his masterly analysis of the mental and physical spheres of war, neglected the material-man's tools. If he thereby ensured to his work an enduring permanence, he also, if unwittingly, ensured permanent injury to subsequent generations who allowed themselves to forget that the spirit cannot win battles when the body has been killed through failure to provide it with up-to-date weapons.

- B.H. Liddell Hart (Thoughts on War)

✧✧✧

Of what use is decisive victory in battle if we bleed to death as a result of it?

- B.H. Liddell Hart (Paris or the Future of War)

✧✧✧

The wider role of mobility and offensive power lies in the air. And the air force appears to be cast for the decisive role, as the heirs of Alexander's "Companion" cavalry. Thus, as of old the forces of a nation for war on land were thought of in terms of infantry and cavalry, though each had its several sub-divisions, so in the future we need to think of the army and the air force as the two main components of military power.

- B.H. Liddell Hart (Thoughts on War)

✧✧✧

A general's dispatch should have more honesty than a politician's electioneering pamphlets.

- B.H. Liddell Hart (Thoughts on War)

✧✧✧

For whoever habitually suppresses truth in the interests of tact will produce a deformity from the womb of his thought.

- B.H. Liddell Hart (How to Win Wars)

✧✧✧

I have long come to realize that the type of senior officer who is always preaching loyalty essentially means loyalty, on the part of his subordinates, to his own interests.

- B.H. Liddell Hart, 1932

✧✧✧

Movement is the safety-valve of fear.

- B.H. Liddell Hart (Thoughts on War)

✧✧✧

The argument that "defence always catches up with attack" is much favoured by those who are opposed to progressive developments, even in the means of defence. History justifies it to a certain extent. But the assertion overlooks the historical fact that defence has often failed to catch up in time to save armies and navies, and the countries they are intended to protect, from defeat in the interval.

- B.H. Liddell Hart (Thoughts on War)

✧✧✧

Originality is the most vital of all military virtues, as two thousand years of wars attest. In peace it is at a discount, for it causes the disturbance of comfortable ways without producing dividends, as in civil life. But in war originality bears a higher premium than it can ever do in a civil profession. For its application can overthrow a nation and change the course of history in the proverbial twinkling of an eye.

- B.H. Liddell Hart (Thoughts on War)

✧✧✧

It is ever a paradox in military affairs that the only way to obtain license for intellectual ideas is to prove oneself an expert in conventional practices.

- B.H. Liddell Hart (Thoughts on War)

✧✧✧

Air forces can be switched from one objective to another. They are not committed to any one course of action as an army is, by its bulk, complexity, and relatively low mobility. While their action should be concentrated, it can be quickly concentrated afresh against other objectives, not only in a different place, but of a different kind.

- B.H. Liddell Hart (Thoughts on War)

✧✧✧

So why not coin a new principle for the new force-*fluid, or distributed, concentration.* To strike, by fire alone, at the greatest number of points in the shortest time over the widest area. And without ever making contact in the present tactical sense. Never giving the enemy a target, yet enticing him to waste his ammunition and keeping his nerves at an exhaustingly high tension.

- B.H. Liddell Hart (Thoughts on War)

✧✧✧

Under the new conditions of warfare it may happen that the *cumulative* effect of partial success, or even mere threat, at a number of points may be greater than the effect of complete success at one point.

- B.H. Liddell Hart (How to Win Wars)

✧✧✧

To disarm is more potent than to kill.

- B.H. Liddell Hart (Thoughts on War)

✧✧✧

The most potent action of mechanized forces will be to supplement their own air force in interrupting the enemy's "circulation."

- B.H. Liddell Hart (Thoughts on War)

✧✧✧

Thus progressive butchery, politely called "attrition," becomes the essence of war. To kill, if possible, more of the enemy troops than your own side loses, is the sum total of this military creed, which attained its tragi-comic climax on the Western front in the Great War.

- B.H. Liddell Hart (Thoughts on War)

✧✧✧

In reality, it is more fruitful to wound than to kill. While the dead man lies still, counting only one man less, the wounded man is a progressive drain upon his side.

- B.H. Liddell Hart (Thoughts on War)

✧✧✧

While a stroke close to the rear of the opposing army is apt to have more effect on the minds of the enemy's troops, a stroke farther back tends to have more effect on the mind of the enemy commander-and it is in the minds of commanders that the issue of battles is really decided.

- B.H. Liddell Hart (Thoughts on War)

✧✧✧

Paralysis, rather than destruction, is the true aim in war, and the most far-reaching in its effects.

- B.H. Liddell Hart (Thoughts on War)

✧✧✧

The object of the rear attack is not itself to crush the enemy, but to unhinge his morale and dispositions so that his dislocation renders the subsequent delivery of a decisive blow both practicable and easy.

- B.H. Liddell Hart (Thoughts on War)

✧✧✧

As when snow is squeezed into a snowball, direct pressure has always the tendency to harden and consolidate the resistance of an opponent-and the more compact it becomes the more difficult it is to melt.

- B.H. Liddell Hart (Thoughts on War)

✧✧✧

A commander should have a profound understanding of human nature, the knack of smoothing out troubles, the power of winning affection while communicating energy, and the capacity for ruthless determination where require by circumstances. He needs to generate an electrifying current, and to keep a cool head in applying it.

- Capt. Sir Basil Liddell Hart

✧✧✧

Air Power is, above all, a psychological weapon—and only short-sighted soldiers, too battle-minded, underrate the importance of psychological factors in war.

- Sir Basil H. Liddell-Hart

✧✧✧

A modern state is such a complex and interdependent fabric that it offers a target highly sensitive to a sudden and overwhelming blow from the air.

- B. H. Liddell-Hart

✧✧✧

The spirit of discipline, as distinct from its outward and visible guises, is the result of association with martial traditions and their living embodiment.

- Sir Basil H. Liddell-Hart , Thoughts on War, 1944

✧✧✧

There are over two thousand years of experience to tell us that the only thing harder than getting a new idea into the military mind is to get an old one out.

- Sir Basil Liddell Hart

✧✧✧

It is a mistake to assume that every unit officer will make all that there is to be made out of his situation; most of them succumb to a certain inertia. Then it is simply reported that for some reason or another, this or that cannot be done—reasons are always easy enough to think up. People of this kind must be made to feel the authority of the commander and be shaken out of their apathy.

- Sir Basil Liddell Hart (The Rommel Papers)

✧✧✧

Wars can be won or lost before the fighting begins.

- Sir Basil Liddell Hart

✧✧✧

Military education hitherto has not been designed to teach a scientific approach to problems, but rather to develop executive skill and foster the spirit of loyalty.

- Sir Basil Liddell Hart

✧✧✧

The hydrogen bomb is not the answer to the Western peoples' dream of full and final insurance of their security ... While it has increased their striking power it has sharpened their anxiety and deepened their sense of insecurity.

- Sir Basil H. Liddell-Hart

✧✧✧

In war the chief incalculable is the human will, which manifests itself in resistance, which in turn lies in the province of tactics. Strategy has not to overcome resistance, except from nature. Its purpose is to diminish the possibility of resistance, and it seeks to fulfil this purpose by exploiting the elements of movement and surprise.

- Sir Basil H. Liddell-Hart

✧✧✧

For even the best of peace training is more theoretical than practical experience ... indirect practical experience may be the more valuable because infinitely wider.

- Sir Basil H. Liddell-Hart

✧✧✧

... the predominance of moral factors in all military decisions. On them constantly turns the issue of war and battle. In the history of war they form the more constant factors, changing only in degree, whereas the physical factors are different in almost every war and every military situation.

- Sir Basil H. Liddell-Hart

✧✧✧

The most consistently successful commanders, when faced by an enemy in a position that was strong naturally or materially, have hardly ever tackled it in a direct way. And when, under pressure of circumstances, they have risked a direct attack, the result has commonly been to blot their record with a failure.

- Sir Basil H. Liddell-Hart

✧✧✧

Natural hazards, however formidable, are inherently less dangerous and less uncertain than fighting hazards. All conditions are more calculable, all obstacles more surmountable than those of human resistance.

- Sir Basil H. Liddell-Hart

✧✧✧

The practical value of history is to throw the film of the past through the material projector of the present on to the screen of the future.

- Sir Basil H. Liddell-Hart

✧✧✧

Science and technology have produced a greater transformation of the physical conditions and apparatus of life in the past hundred years than had taken place in the previous two thousand years. Yet when men turn these tremendous new powers to a war purpose, they employ them as recklessly as their ancestors employed the primitive means of the past, and they pursue the same traditional ends without regard to the difference of effect.

- Sir Basil H. Liddell-Hart (1895-1970)
Why don't we learn from history?

✧✧✧

History is a universal experience-infinitely longer, wider, and more varied than any individual's experience. How often do we hear people claim knowledge of the world and of life because they are sixty or seventy years old? . . There is no excuse for any literate person if he is less than three thousand years old in mind.

- Sir Basil H. Liddell-Hart

✧✧✧

History is a catalogue of mistakes. It is our duty to profit by them.

- Sir Basil H. Liddell-Hart

✧✧✧

The most effective indirect approach is one that lures or startles the opponent into a false move—so that, as in ju-jitsu, his own effort is turned into the lever of his overthrow.

- Sir Basil H. Liddell-Hart

✧✧✧

The downfall of civilized states tends to come not from the direct assaults of foes, but from internal decay combined with the consequences of exhaustion in war.

- Sir Basil H. Liddell-Hart

✧✧✧

In a campaign against more than one state or army, it is more fruitful to concentrate first against the weaker partner than to attempt the overthrow of the stronger in the belief that the latter's defeat will automatically involve the collapse of the others.

- Sir Basil H. Liddell-Hart

✧✧✧

While hitting one must guard ... In order to hit with effect, the enemy must be taken off his guard.

- Sir Basil H. Liddell-Hart

✧✧✧

For whoever habitually suppresses the truth in the interests of tact will produce a deformity from the womb of his thought.

- Sir Basil H. Liddell-Hart

✧✧✧

The military weapon is but one of the means that serve the purposes of war: one out of the assortment which grand strategy can employ.

- Sir Basil H. Liddell-Hart

✧✧✧

While there are many causes for which a state goes to war, its fundamental object can be epitomized as that of ensuring the continuance of its policy—in face of the determination of the opposing state to pursue a contrary policy. In the human will lies the source and mainspring of conflict.

- Sir Basil H. Liddell-Hart

✧✧✧

It is thus more potent, as well as more economical, to disarm the enemy than to attempt his destruction by hard fighting ... A strategist should think in terms of paralysing, not of killing.

- Sir Basil H. Liddell-Hart

✧✧✧

While the nominal strength of a country is represented by its numbers and resources, this muscular development is dependent on the state of its internal organs and nerve-system—upon its stability of control, morale, and supply.

- Sir Basil H. Liddell-Hart

✧✧✧

To ensure attaining an objective, one should have alternate objectives. An attack that converges on one point should threaten, and be able to diverge against another. Only by this flexibility of aim can strategy be attuned to the uncertainty of war.

- Sir Basil H. Liddell-Hart

✧✧✧

The higher level of grand strategy [is] that of conducting war with a far-sighted regard to the state of the peace that will follow.

- Sir Basil H. Liddell-Hart

✧✧✧

The nearer the cutting off point lies to the main force of the enemy, the more *immediate* the effect; whereas the closer to the strategic base it takes place, the *greater* the effect.

- Sir Basil H. Liddell-Hart

✧✧✧

As has happened so often in history, victory had bred a complacency and fostered an orthodoxy which led to defeat in the next war.

- Sir Basil H. Liddell-Hart

(Strategy, 1954; on the French army between the World Wars)

✧✧✧

This high proportion of history's decisive campaigns, the significance of which is enhanced by the comparative rarity of the direct approach, enforces the conclusion that the indirect is by far the most hopeful and economic form of strategy.

- Sir Basil H. Liddell-Hart

✧✧✧

The more closely [the German army] converged on [Stalingrad], the narrower became their scope for tactical manoeuvre as a lever in loosening resistance. By contrast, the narrowing of the frontage made it easier for the defender to switch his local reserves to any threatened point on the defensive arc.

- Sir Basil H. Liddell-Hart

✧✧✧

Their strength became split in diverging directions—due partly to divided minds at the top, but also, ironically, to dazzling initial success in all directions. Instead of keeping a single line of operation that threatened alternate objectives, they were led to pursue several lines of operation, each too obviously aiming at a single objective, which thus became easier for the defender to cover. Moreover, in each case the attacker's direction became obvious at the same time that his drive was becoming a precarious stretch of his own supply line.

- Sir Basil H. Liddell-Hart
(Strategy, 1954; on German failure in WWII)

✧✧✧

[the blurring of the line between policy and strategy] encouraged soldiers to make the preposterous claim that policy should be subservient to their conduct of operations, and (especially in democratic countries) it drew the statesman on to overstep the definite border of his sphere and interfere with his military employees in the actual use of their tools.

- Sir Basil H. Liddell-Hart

✧✧✧

The more usual reason for adopting a strategy of limited aim is that of awaiting a change in the balance of force ... The essential condition of such a strategy is that the drain on him should be disproportionately greater than on oneself.

- Sir Basil H. Liddell-Hart

✧✧✧

To foster the people's willing spirit is often as important as to possess the more concrete forms of power.

- Sir Basil H. Liddell-Hart

✧✧✧

The effect to be sought is the dislocation of the opponent's mind and dispositions—such an effect is the true gauge of an indirect approach.

- Sir Basil H. Liddell-Hart

✧✧✧

No man can exactly calculate the capacity of human genius and stupidity, nor the incapacity of will.

- Sir Basil H. Liddell-Hart

✧✧✧

In the case of a state that is seeking not conquest but the maintenance of its security, the aim is fulfilled if the threat is removed—if the enemy is led to abandon his purpose.

- Sir Basil H. Liddell-Hart

✧✧✧

Direct pressure always tends to harden and consolidate the resistance of an opponent.

- Sir Basil H. Liddell-Hart

✧✧✧

[The] aim is not so much to seek battle as to seek a strategic situation so advantageous that if it does not of itself produce the decision, its continuation by a battle is sure to achieve this. In other words, dislocation is the aim of strategy.

- Sir Basil H. Liddell-Hart

✧✧✧

For if we merely take what obviously appears the line of least resistance, its obviousness will appeal to the opponent also; and this line may no longer be that of least resistance. In studying the physical aspect, we must never lose sight of the psychological, and only when both are combined is the strategy truly an indirect approach, calculated to dislocate the opponent's balance.

- Sir Basil H. Liddell-Hart

✧✧✧

It is folly to imagine that the aggressive types, whether individuals or nations, can be bought off ... since the payment of danegeld stimulates a demand for more danegeld. But they can be curbed. Their very belief in force makes them more susceptible to the deterrent effect of a formidable opposing force.

- Sir Basil H. Liddell-Hart

(Note: Danegeld was an annual land tax in England, in the 10th and 11th centuries, imposed to provide funds for protection against the Danes)

✧✧✧

An army should always be so distributed that its parts can aid each other and combine to produce the maximum possible concentration of force at one place, while the minimum force necessary is used elsewhere to prepare the success of the concentration.

- Sir Basil H. Liddell-Hart

✧✧✧

In any problem where an opposing force exists and cannot be regulated, one must foresee and provide for alternative courses. Adaptability is the law which governs survival in war as in life ... To be practical, any plan must take account of the enemy's power to frustrate it; the best chance of overcoming such obstruction is to have a plan that can be easily varied to fit the circumstances met;

- Sir Basil H. Liddell-Hart

✧✧✧

The unexpected cannot guarantee success, but it guarantees the best chance of success.

- Sir Basil H. Liddell-Hart

✧✧✧

Air forces offered the possibility of striking a the enemy's economic and moral centres without having first to achieve 'the destruction of the enemy's main forces on the battlefield'. Air-power might attain a direct end by indirect means—hopping over opposition instead of overthrowing it.

- Sir Basil H. Liddell-Hart

✧✧✧

It should be the aim of grand strategy to discover and pierce the Achilles' heel of the opposing government's power to make war. Strategy, in turn, should seek to penetrate a joint in the harness of the opposing forces. To apply one's strength where the opponent is strong weakens oneself disproportionately to the effect attained. To strike with strong effect, one must strike at weakness.

- Sir Basil H. Liddell-Hart

✧✧✧

With growing experience, all skilful commanders sought to profit by the power of the defensive, even when on the offensive.

- Sir Basil H. Liddell-Hart

✧✧✧

Inflict the least possible permanent injury, for the enemy of to-day is the customer of the morrow and the ally of the future.

- Sir Basil H. Liddell-Hart

✧✧✧

If you find your opponent in a strong position costly to force, you should leave him a line of retreat as the quickest way of loosening his resistance. It should, equally, be a principle of policy, especially in war, to provide your opponent with a ladder by which he can climb down.

- Sir Basil H. Liddell-Hart

✧✧✧

It is only to clear from history that states rarely keep faith with each other, save in so far (and so long) as their promises seem to them to combine with their interests.

- Sir Basil H. Liddell-Hart

✧✧✧

The implied threat of using nuclear weapons to curb guerrillas was as absurd as to talk of using a sledge hammer to ward off a swarm of mosquitoes.

- Sir Basil H. Liddell-Hart

✧✧✧

The aim of the Strategy of Indirect Approach is not so much to seek battle as to seek a strategic situation so advantageous that if it does not of itself produce the decision, its continuation by a battle is sure to achieve this.

- Sir Basil H. Liddell-Hart

✧✧✧

War is always a matter of doing evil in the hope that good may come of it.

- Sir Basil H. Liddell-Hart

✧✧✧

Throughout the ages, effective results in war have been rarely attained unless the approach has had such indirectness as to ensure the opponents unreadiness to meet it. The indirectness has usually been physical, and always psychological. In strategy, the longest way around is often the shortest way home".

- Basil Liddell Hart, Strategy,

✧✧✧

The weapon where the man is sitting in is always superior against the other.

- Colonel Erich 'Bubi' Hartmann, German Air Force

✧✧✧

He lined up his disreputable paladins in the darkness, and spoke—

"Sergeant Mc'Nab, how many men are present?"

"Eighteen, Sir."

The platoon had gone into action thirty-four strong.

"How many men are deficient of an emergency ration? I can make a good guess, but you had better find out."

Five minutes later the Sergeant reported. Cockerell's guess was correct. The British private has only one point of view about the portable property of the State. To him, as an individual, the sacred emergency ration is an unnecessary encumbrance, and the carrying thereof a "fatigue." Consequently, when engaged in battle one of the first (of many) things which he jettisons is this very ration. When all is over, he reports with unctuous solemnity that the provender in question has been blown out of his haversack by a shell. The Quartermaster-Sergeant writes it off as "lost owing to the exigencies of military service," and indents for another.

- Ian Hay, All In It; "K(1)" Carries On, 1917

✧✧✧

Without a theory, the facts are silent.

- Friederich A. von Hayek

✧✧✧

The mind is a not vessel to be filled, but a lamp to be lighted.

- Hebrew proverb

✧✧✧

History has a certain cunningness and history is always rational.

- Georg Wilhelm Friedrich Hegel

✧✧✧

There was only one catch and that was Catch-22, which specified that a concern for one's own safety in the face of dangers that were real and immediate was the process of a rational mind. Orr was crazy and could be grounded. All he had to do was ask; and as soon as he did, he would no longer be crazy and would have to fly more missions. Orr would be crazy to fly more missions and sane if he didn't, but if he was sane, he had to fly them. If he flew them, he was crazy and didn't have to; but if he didn't want to, he was sane and had to. Yossarian was moved very deeply by the absolute simplicity of this clause of Catch-22 and let out a respectful whistle.

- Joseph Heller

✧✧✧

"What is the moral difference, if any, between the soldier and the civilian?" "The difference," I answered carefully, "lies in the field of civic virtue. A soldier accepts personal responsibility for the safety of the body politic of which he is a member, defending it, if need be, with his life. The civilian does not."

- Robert A. Heinlein, Starship Troopers, 1959

✧✧✧

Peace is an extension of war by political means.

- Robert Heinlein

✧✧✧

Always keep your clothes and your weapons where you can find them in the dark.

- Robert Heinlein

✧✧✧

There is a limit on our ability to observe reality with precision.

- Heisenberg's Uncertainty Principle

✧✧✧

Discipline is simply the art of making the soldiers fear their officers more than the enemy.

- Helvetius (1715 -1771), French philosopher

✧✧✧

We have no freedom of choice between good and evil. There is no such thing as absolute right—ideas of justice and injustice change according to customs.

- Helvetius

✧✧✧

Education made us what we are.

- Helvetius

✧✧✧

Once we have a war there is only one thing to do. It must be won. For defeat brings worse things than any that can ever happen in war.

- Ernest Miller Hemmingway

✧✧✧

For a war to be just three conditions are necessary—public authority, just cause, right motive.

- Ernest Miller Hemmingway

✧✧✧

Defence is the stronger form with the negative object, and attack the weaker form with the positive object.

- Ernest Miller Hemmingway

✧✧✧

I know war as few other men now living know it, and nothing to me is more revolting. I have long advocated its complete abolition, as its very destructiveness on both friend and foe has rendered it useless as a method of settling international disputes.

- Ernest Miller Hemmingway

✧✧✧

They wrote in the old days that it is sweet and fitting to die for one's country. But in modern war, there is nothing sweet nor fitting in your dying. You will die like a dog for no good reason.

- Ernest Hemmingway

✧✧✧

Wars are caused by undefended wealth.

- Ernest Miller Hemmingway

✧✧✧

The sinews of war are five—men, money, materials, maintenance (food) and morale.

- Ernest Miller Hemmingway

✧✧✧

Never think that war, no matter how necessary, nor how justified, is not a crime.

- Ernest Miller Hemmingway

✧✧✧

In modern war... you will die like a dog for no good reason.

- Ernest Miller Hemmingway

✧✧✧

Once we have a war there is only one thing to do. It must be won. For defeat brings worse things than any that can ever happen in war.

- Ernest Miller Hemingway

✧✧✧

The highest generalship is to compel the enemy to disperse his army, and then to concentrate superior force against each fraction in turn.

- Colonel GFR Henderson (1854-1903)

✧✧✧

The rules of strategy are few and simple. They may be learned in a week. They may be taught by familiar illustrations or a dozen diagrams. But such knowledge will no more teach a man to lead an army like Napoleon than a knowledge of grammar will teach him to write like Gibbon.

- Colonel GFR Henderson

✧✧✧

The line of supply may be said to be as vital to the existence of an army as the heart to the life of a human being. Just as the duellist who finds his adversary's point menacing him with certain death, and his own guard astray, is compelled to conform to his adversary's movements, and to content himself with warding off his thrusts, so the commander whose communications are suddenly threatened finds himself in a false position, and he will be fortunate if he has not to change all his plans, to split up his force into more or less isolated detachments, and to fight with inferior numbers on ground which he has not had time to prepare, and where defeat will not be an ordinary failure, but will entail the ruin or surrender of his whole army.

- Colonel GFR Henderson

✧✧✧

INVICTUS

Out of the night that covers me,
Black as the pit from pole to pole,
I thank whatever gods may be
For my unconquerable soul.

In the fell clutch of circumstance
I have not winced nor cried aloud.
Under the bludgeonings of chance
My head is bloody, but unbowed.

Beyond this place of wrath and tears
Looms but the Horror of the shade,
And yet the menace of the years
Finds and shall find me unafraid.

It matters not how strait the gate,
How charged with punishments the scroll,
I am the master of my fate:
I am the captain of my soul.

- William Ernest Henley (1849-1903)
Disambiguation

✧✧✧

The battle, sir, is not to the strong alone; it is to the vigilant, the active, the brave...

- Patrick Henry

✧✧✧

It is in vain, sir, to extenuate the matter. Gentlemen may cry, Peace, Peace—but there is no peace. The war is actually begun! The next gale that sweeps from the north will bring to our ears the clash of resounding arms! Our brethren are already in the field! Why stand we here idle? What is it that gentlemen wish? What would they have? Is life so dear, or peace so sweet, as to be purchased at the price of chains and slavery? Forbid it, Almighty God! I know not what course others may take; but as for me, give me liberty or give me death!

- Patrick Henry, 23 March1775

✧✧✧

Out of every 100 men, ten shouldn't even be there, Eighty are just targets, Nine are the real fighters, and we are lucky to have them, for they make the battle. Ah, but the one, One is a warrior, and he will bring the others back.

- Heraclitus

✧✧✧

Character is destiny.

- Heraclitus

✧✧✧

No man ever steps in the same river twice, for it's not the same river and he's not the same man.

- Heraclitus

✧✧✧

Nothing endures but change.

- Heraclitus

✧✧✧

Much learning does not teach understanding.

- Heraclitus

✧✧✧

War is the mother of everything.

- Heraclitus

✧✧✧

Circumstances rule men; men do not rule circumstances.

- Herodotus, Greek historian c. 484-425 B.C.

✧✧✧

Force has no place where there is need of skill.

- Herodotus

✧✧✧

Men trust their ears less than their eyes.

- Herodotus

✧✧✧

Some men give up their designs when they have almost reached the goal, while others, on the contrary, obtain a victory by exerting, at the last moment, more vigorous efforts than ever before.

- Herodotus (The Histories)

✧✧✧

"He asked, 'Croesus, who told you to attack my land and meet me as an enemy instead of a friend?'

The King replied, 'It was caused by your good fate and my bad fate. It was the fault of the Greek gods, who with their arrogance, encouraged me to march onto your lands. Nobody is mad enough to choose war whilst there is peace. During times of peace, the sons bury their fathers, but in war it is the fathers who send their sons to the grave."

- Herodotus

✧✧✧

It is the greatest houses and the tallest trees that the gods bring low with bolts and thunder. For the gods love to thwart whatever is greater than the rest. They do not suffer pride in anyone but themselves.

- Herodotus

✧✧✧

All men, if asked to choose the best way of ordering life would choose their own.

- Herodotus, Greek historian, 474 BC-424 BC

✧✧✧

He is the best man who, when making his plans, fears and reflects on everything that can happen to him, but in the moment of action is bold.

- Herodotus

✧✧✧

The destiny of man is in his own soul.

- Herodotus

✧✧✧

A war put off is not a war avoided.

- Charlton Heston

✧✧✧

The enemy bombards our front not only with a drumfire of artillery, but also with a drumfire of printed paper. Besides bombs, which kill the body, his airmen also throw down leaflets which are intended to kill the soul.

- Field Marshal Paul Von Hindenburg, 1847-1934

✧✧✧

All we know is that, at times, fighting the Russians, we had to remove the piles of enemy bodies from before our trenches, so as to get a clear field of fire against new waves of assault.

- Paul Von Hindenburg

✧✧✧

In war, only the simple succeeds.

- Paul Von Hindenburg

✧✧✧

However, the fact that the tanks had now been raised to such a pitch of technical perfection that they could cross our undamaged trenches and obstacles did not fail to have a marked effect on our troops.

- Paul Von Hindenburg

✧✧✧

The charges that Germany is guilty of the greatest of all wars, we, the German people, repudiate in all its phases. Not envy, hate, nor eagerness for conquest caused us to resort to weapons. War was a last resort for us, and the requiring of the greatest sacrifices of the entire people was the last means of maintaining our prestige against a host of enemies. With pure hearts we marched out to defend the Fatherland, and with pure hands the German Army wielded the sword. Germany is ever ready to prove it before impartial judges.

- Paul Von Hindenburg

✧✧✧

Erect and martial, President Generalfeldmarschall Paul Ludwig Hans von Beneckendorff und von Hindenburg arrived at Tannenberg, East Prussia, to unveil a War memorial to the soldiers who fell in the historic Battle of Tannenberg, fought from 26 to 31 August 1914. In it General-Oberst von Hindenburg, as he was then, and Major General von Ludendorff achieved their greatness in Germany by destroying the Russian Second Army Corps, commanded by famed General Samsonoff.

More than 100,000 people gathered to witness the ceremony. Six miles of veterans lined up to do honour to their old military chief. Some of them were dressed in field grey; others were resplendent in plumed helmets and gold-braided tunics of imperial days. Gathered there, too, were many of the highest Reich authorities, from Chancellor William Marx and several of his cabinet to Marshal von Mackensen, Generals von Francois and von Ludendorff:

Clad in his marshal's uniform, with the baton of his rank in his left hand, the aged Hindenburg, almost 80, passed through the cheering throng, stopping now and then to say a few words to a former comrade-in-arms. Grim, cool, calm, yet genial enough on occasion, Germans recall a story about their President that exemplifies his peculiar wit:

One of his old friends is alleged to have asked him:
"What do you do when you get excited?"
"I whistle," replied the President.
"But I have never heard you whistle."
"I never have."

- Paul Von Hindenburg

✧✧✧

... General Headquarters decided that it was advisable to have pass-words for fighting troops, so every twenty-four hours a pass-word was issued from Brigade, who selected it. The new word commenced from the evening "Stand to," and our Brigade, the 73rd, chose the names of the officers commanding battalions at first -Greene, Mobbs Murphy, etc. On the fifth and following nights, ordinary words such as "Rabbit," "Apple," etc., were introduced.

Some of the "old toughs," however, found some difficulty in remembering the absurd words which followed, and on one of the dark nights of this tour, seeing a man approaching me, I called out, "Halt, who goes there?" only to get the following unusual reply, "Begad, I was a rabbit last night, a spud the night afore, and I'm damned if I know what I'm meant to be at all to-night."

It was 8645 Pte Corbally. He apologised profusely when he recognised me. I told him the pass-word, and went on my tour laughing. Corbally was a treasure.

- Captain F.C. Hitchcock, M.C.,
"Stand To" A Diary of the Trenches, 1915- 18, 1937

✧✧✧

We have resolved to endure the unendurable and suffer what is insufferable.

- Emperor Hirohito

✧✧✧

Terrorism is the tactic of demanding the impossible, and demanding it at gunpoint.

- Christopher Hitchens, "Terrorism: Notes Toward a Definition"

✧✧✧

It must be thoroughly understood that the lost land will never be won back by solemn appeals to the God, nor by hopes in any League of Nations, but only by the force of arms.

- Adolf Hitler

✧✧✧

How to achieve the moral breakdown of the enemy before the war has started—that is the problem that interests me. Whoever has experienced war at the front will want to refrain from all avoidable bloodshed.

*- Adolf Hitler (*from *Hitler Speaks)*

✧✧✧

I will not wage war against women and children! I have instructed my air force to limit their attacks to military objectives. However, if the enemy should conclude from this that he might get away with waging war in a different manner he will receive an answer that he'll be knocked out of his wits!

- Adolf Hitler, speech before the Reichstag, 1 September 1939.

✧✧✧

The conviction of the justification of using even the most brutal weapons is always dependent on the presence of a fanatical belief in the necessity of the victory of a revolutionary new order on this globe.

- Adolf Hitler, 'Mein Kampf.'

✧✧✧

A single blow must destroy the enemy... without regard of losses... a gigantic all-destroying blow.

- Adolf Hitler

✧✧✧

Become strong again in spirit, strong in will, strong in endurance, strong to bear all sacrifices.

- Adolf Hitler

✧✧✧

Strength lies not in defence but in attack.

- Adolf Hitler

✧✧✧

Words build bridges into unexplored regions.

- Adolf Hitler

✧✧✧

Great liars are also great magicians.

- Adolf Hitler

✧✧✧

People have killed only when they could not achieve their aim in other ways ... there is a broadened strategy, with intellectual weapons ... why should I demoralize the enemy by military means if I can do so better and more cheaply in other ways?

- Adolf Hitler

✧✧✧

Demoralize the enemy from within by surprise, terror, sabotage, assassination. This is the war of the future.

- Adolf Hitler

✧✧✧

Let us never forget the duty, which we have taken upon us.

- Adolf Hitler

✧✧✧

Just as the world cannot live on wars, so people cannot on revolutions.

- Adolf Hitler

✧✧✧

Our strategy is to destroy the enemy from within, to conquer him through himself.

- Adolf Hitler

✧✧✧

Germany is prepared to agree to any solemn pact of non-aggression, because she does not think of attacking but only acquiring security.

- Adolf Hitler (1933)

✧✧✧

We are all proud that through God's powerful aid, we have become once more true Germans.

- Adolf Hitler

✧✧✧

National Socialist Germany wants peace because of its fundamental convictions. And it wants peace also owing to the realization of the simple primitive fact that no war would be likely essentially to alter the distress in Europe... The principal effect of every war is to destroy the flower of the nation... Germany needs peace and desires peace!

- Adolf Hitler, 21 May 1935

✧✧✧

This year will go down in history. For the first time, a civilized nation has full gun registration! Our streets will be safer, our police more efficient, and the world will follow our lead into the future!

- Adolf Hitler, 1935
on the passing of the German Weapons Act.

✧✧✧

England, unlike in 1914, will not allow herself to blunder into a war lasting for years.... Such is the fate of rich countries.. .Not even England has the money nowadays to fight a world war. What should England fight for? You don't get yourself killed over an ally.

- Adolf Hitler (1939)

✧✧✧

The personification of the devil as the symbol of all evil assumes the living shape of the Jew.

- Adolf Hitler

✧✧✧

The war against Russia will be such that it cannot be conducted in a knightly fashion. This struggle is one of ideologies and racial differences and will have to be conducted with unprecedented, unmerciful and unrelenting harshness.

- Adolf Hitler (1941)

✧✧✧

To be a leader means to be able to move masses.

- Adolf Hitler

✧✧✧

Whatever goal, man has reached is due to his originality plus his brutality.

- Adolf Hitler

✧✧✧

If we had these rockets in 1939, we should never have had this war.

- Adolf Hitler, regards the V-1 rocket.

✧✧✧

We will not capitulate—no, never! We may be destroyed, but if we are, we shall drag a world with us—a world in flames.

- Adolf Hitler

✧✧✧

Struggle is the father of all things. It is not by the principles of humanity that man lives or is able to preserve himself above the animal world, but solely by means of the most brutal struggle. If you do not fight, life will never be won.

- Adolf Hitler

✧✧✧

There could be no issue between the Church and the State. The Church, as such, has nothing to do with political affairs. On the other hand, the State has nothing to do with the faith or inner organization of the Church.

- Adolf Hitler

✧✧✧

The man who has no sense of history, is like a man who has no ears or eyes.

- Adolf Hitler

✧✧✧

It is always more difficult to fight against faith than against knowledge.

- Adolf Hitler

✧✧✧

Man has become great through struggle.

- Adolf Hitler

✧✧✧

The German people are not a warlike nation. It is a soldierly one, which means it does not want a war, but does not fear it. It loves peace but also loves its honour and freedom.

- Adolf Hitler

✧✧✧

Always before God and the world, the stronger has the right to carry through what he wills.

- Adolf Hitler

✧✧✧

The majority can never replace the man.

- Adolf Hitler

✧✧✧

What we have to fight for is the freedom and independence of the fatherland, so that our people may be enabled to fulfil the mission assigned to it by the creator.

- Adolf Hitler

✧✧✧

I will never allow anyone to divide this people once more into religious camps, each fighting the other.

- Adolf Hitler

✧✧✧

Struggle is the father of all things, virtue lies in blood, leadership is primary and decisive.

- Adolf Hitler

✧✧✧

The world will not help, the people must help themselves. Its own strength is the source of life. That strength the Almighty has given us to use; that in it and through it, we may wage the battle of our life The others in the past years have not had the blessing of the Almighty—of Him who in the last resort, whatever man may do, holds in His hands the final decision.

Lord God, let us never hesitate or play the coward.

- Adolf Hitler

✧✧✧

Only force rules. Force is the first law.

- Adolf Hitler

✧✧✧

The only people I have been able to use are those who fought.

- Adolf Hitler

✧✧✧

The broad masses of a population are more amenable to the appeal of rhetoric than to any other force.

- Adolf Hitler

✧✧✧

The leader of genius must have the ability to make different opponents appear as if they belonged to one category.

- Adolf Hitler

✧✧✧

The victor will never be asked if he told the truth.

- Adolf Hitler

✧✧✧

When an opponent declares, I will not come over to your side, I calmly say, Your child belongs to us already... What are you? You will pass on. Your descendants, however, now stand in the new camp. In a short time they will know nothing else but this new community.

- Adolf Hitler

✧✧✧

Who says I am not under the special protection of God?

- Adolf Hitler

✧✧✧

War between China and Taiwan is unthinkable today. It makes no sense. It is as unthinkable as an Iraqi invasion of Kuwait was in July 1990, as unthinkable as China entering the Korean War against the United States was in November 1950, as unthinkable as Britain having to expel the Argentines from the Falklands seemed in 1982.

- Foreign affairs analyst Jim Hoagland
11 February 1996, Washington Post

✧✧✧

Force and fraud are in war the two cardinal virtues.

- Thomas Hobbes

✧✧✧

Small minds serve the letter of the law.
Great minds serve justice.

-Thomas Hobbes

✧✧✧

I believe in compulsory cannibalism. If people were forced to eat what they killed, there would be no more wars.

- Abbie Hoffman

✧✧✧

Of all the disasters of Vietnam the worst could be our unwillingness to learn enough from them.

- Stanley Hoffman

✧✧✧

Sometimes it is entirely appropriate to kill a fly with a sledge hammer.

- Major Holdridge

✧✧✧

Older men declare war. But it is youth that must fight and die. And it is youth who must inherit the tribulation, the sorrow, and the triumphs that are the aftermath of war.

- Herbert Hoover, speech, 1944

✧✧✧

About the time we can make the ends meet, somebody moves the ends.

- Herbert Hoover, President of the United States (1929–1933)

✧✧✧

A ship in port is safe, but that's not what ships are built for.

- Rear Admiral Grace Murray Hopper

✧✧✧

If it's a good idea, go ahead and do it. It's much easier to apologize than it is to get permission.

- Rear Admiral Grace Murray Hopper

✧✧✧

Any damn fool can write a plan. It's the execution that gets all screwed up.

- Lt General James Hollingworth, US Army

✧✧✧

Yes, we love peace, but we are not willing to take wounds for it, as we are for war.

- John Andrew Holmes

✧✧✧

One line [of Montgomery's] which made the greatest impact with infantrymen would go as follows:

'You. What's your most valuable possession?' 'My rifle, Sir.'

'No, it isn't; it's your life, and I'm going to save it for you. Now listen to me...'

He would then go on to explain how he would never make the infantry attack without full artillery and air support. Such visits were, as Richard Lamb records, almost invariably: "successful beyond expectation." He had, after all, spent a lifetime with soldiers, and he loved and understood them, especially young men; he could mesmerize them by his speeches, and he revelled in his popularity. Now he found that the average young soldier expected to die in the coming assault, and he sought to dispel their gloom by instilling a revivalist fervour. In this he was right, for morale in Britain was low.

- Alistaire Horne, The Lonely Leader; Monty, 1944-1945

✧✧✧

Progress comes from the intelligent use of experience.

- Elbert Hubbard

✧✧✧

The greatest mistake you can make in this life is to be continually fearing you will make one.

- Elbert Hubbard

✧✧✧

No army can withstand the strength of an idea whose time has come.

- Victor Hugo (1802-1885)
French poet and novelist and dramatist

✧✧✧

Waterloo is a battle of the first rank won by a captain of the second.

- Victor Hugo

✧✧✧

What is history? An echo of the past in the future; a reflex from the future on the past.

- Victor Hugo

✧✧✧

It is my hypothesis that the fundamental source of conflict in this new world will not be primarily ideological or primarily economic. The great divisions among humankind and the dominating source of conflict will be cultural. Nation-states will remain the most powerful actors in world affairs, but the principal conflicts of global politics will occur between nations and groups of different civilizations. The clash of civilizations will dominate global politics. The fault lines between civilizations will be the battle lines of the future.

The West won the world not by the superiority of its ideas or values or religion, but rather by its superiority in applying organized violence. Westerners often forget this fact, non-Westerners never do.

- Samuel P. Huntington
The Clash of Civilizations and the Remaking of World Order

✧✧✧

In Eurasia the great historic fault lines between civilizations are once more aflame. This is particularly true along the boundaries of the crescent-shaped Islamic bloc of nations, from the bulge of Africa to central Asia. Violence also occurs between Muslims, on the one hand, and Orthodox Serbs in the Balkans, Jews in Israel, Hindus in India, Buddhists in Burma and Catholics in the Philippines. Islam has bloody borders.

- Samuel P. Huntington,
The Clash of Civilizations

✧✧✧

Partial truths or half-truths are often more insidious than total falsehoods.

- Samuel P. Huntington

✧✧✧

In the emerging world of ethnic conflict and civilizational clash, Western belief in the universality of Western culture suffers three problems: it is false; it is immoral; and it is dangerous . . . Imperialism is the necessary logical consequence of universalism.

- Samuel P. Huntington
The Clash of Civilizations and the Remaking of World Order

✧✧✧

Islam's borders are bloody and so are its innards. The fundamental problem for the West is not Islamic fundamentalism. It is Islam, a different civilisation whose people are convinced of the superiority of their culture and are obsessed with the inferiority of their power.

- Samuel P. Huntington
The Clash of Civilizations and the Remaking of World Order

✧✧✧

Military society is shaped by a combination of functional imperatives relating to external threats and societal imperatives derived from the nature of society, ... the professional military officer corps is a useful barometer of this change.

- Samuel Huntington
The Soldier and the State

✧✧✧

The military establishment becomes a constabulary force when it is continuously prepared to act, committed to the minimum use of force, and seeks viable international relations, rather than victory because it has incorporated a protective military posture. The constabulary outlook is grounded in, and extends, pragmatic doctrine.

- Samuel Huntington

✧✧✧

I

We've got no place in this outfit for good losers. We want tough hombres who will go in there and win!

- Admiral Jonas Ingram, 1929

✧✧✧

A man may build himself a throne of bayonets, but he cannot sit on it.

- William Ralph Inge

✧✧✧

At a time when the soldier is supplied with an accurate firearm, and when the well-aimed fire of individual men must have more result than ill aimed volleys; when the soldier, in order to fire well and with good effect, must lie comfortably on the ground instead of standing in a close crowded line; when he is, moreover, no longer a mere portion of a stiff machine, since each man can use his weapon with intelligence; when the infantry have ceased to be only food for powder, and have become a combination of single units working independently, at such a time the careful training of the individual soldier must decide the issue of battle.

- Prince Kraft zu Hohenlohe Ingelfingen, Letters on Infantry

✧✧✧

Endow your will with such power. That at every turn of fate it so be, That God Himself asks of His Slave "What is it that pleases thee?

- Allama Iqbal

✧✧✧

War is the ultimate resource of policy, and every nation must be ready in the last instance to protect its vital interests by force of arms unless it is prepared to surrender them to an enemy without a blow.

- Sir Edmund Ironside

✧✧✧

Mrs. Thatcher will now realise that Britain cannot occupy our country and torture our prisoners and shoot our people in their own streets and get away with it. Today we were unlucky, but remember we only have to be lucky once. You will have to be lucky always.

- Provisional Irish Republican Army (IRA) statement

The Brighton hotel bombing occurred on 12 October 1984 at the Grand Hotel in Brighton, England. The bomb was planted by IRA member Patrick Magee. The IRA claimed responsibility and said that it would try again.

✧✧✧

J

The hardships of forced marches are often more painful than the dangers of battle.

- General Thomas Stonewall Jackson (1824–1863)
Confederate General, American Civil War

✧✧✧

Don't say it's impossible! Turn your command over to the next officer. If he can't do it, I'll find someone who can, even if I have to take him from the ranks!

- General Thomas Stonewall Jackson

✧✧✧

When war does come, my advice is to draw the sword and throw away the scabbard.

- General Thomas Stonewall Jackson

✧✧✧

Arms is a profession that, if its principles are adhered to for success, requires an officer do what he fears may be wrong, and yet, according to military experience, must be done, if success is to be attained.
I yield to no man in sympathy for the gallant men under my command; but I am obliged to sweat them tonight, so that I may save their blood tomorrow.

- General Thomas Stonewall Jackson

✧✧✧

Once you get them running, you stay right on top of them, and that way a small force can defeat a large one every time... Only thus can a weaker country cope with a stronger; it must make up in activity what it lacks in strength.

- General Thomas Stonewall Jackson

✧✧✧

Who could not conquer with such troops as these?

- General Thomas Stonewall Jackson

✧✧✧

To move swiftly, strike vigorously, and secure all the fruits of victory, is the secret of successful war.

- General Thomas Stonewall Jackson

✧✧✧

Always mystify, mislead, and surprise the enemy, if possible; and when you strike and overcome him, never let up in the pursuit so long as your men have strength to follow; for an army routed, if hotly pursued, becomes panic-stricken, and can then be destroyed by half their number. The other rule is, never fight against heavy odds, if by any possible manoeuvring you can hurl your own force on only a part, and that the weakest part, of your enemy and crush it. Such tactics will win every time, and a small army may thus destroy a large one in detail, and repeated victory will make it invincible.

- General Thomas Stonewall Jackson

✧✧✧

The only true rule for cavalry is to follow the enemy as long as he retreats.

- General Thomas Stonewall Jackson
to Colonel Munford on 13 June 1862

✧✧✧

Let us cross over the river, and rest under the shade of the trees.

- General Thomas Stonewall Jackson
last words

✧✧✧

The time for war has not yet come, but it will come, and that soon; and when it does come, my advice is to draw the sword and throw away the scabbard.

- General Thomas Stonewall Jackson

✧✧✧

My religious belief teaches me to feel as safe in battle as in bed. God has fixed the time for my death. I do not concern myself about that, but to be always ready, no matter when it may overtake me. ... That is the way all men should live, and then all would be equally brave.

- General Thomas Stonewall Jackson

✧✧✧

Nobody dislikes war more than warriors.

- Daniel James

✧✧✧

The military also serves as an agent of social change. At a minimum, this implies that the army becomes a device for developing a sense of identity—a social psychological element of national unity—which is especially crucial for a nation which has suffered because of colonialism and which is struggling to incorporate diverse ethnic and tribal groups.

-Morris Janowitz
The Military in the Political Development of New Nations

✧✧✧

The military is a social organisation which maintains levels of autonomy while refracting broader social trends.

- Morris Janowitz

✧✧✧

Even in the smallest unit there is an 'iron framework' of organization which serves as a basis of social control. The single concept of military morale must give way, therefore, to a theory of organizational behaviour in which an array of sociological concepts is employed.

- Morris Janowitz

✧✧✧

It appears that a soldier's ability to resist is a function of the capacity of his immediate primary group (his squad or section) to avoid social disintegration. When the individual's immediate group, and its supporting formations, met his basic organic needs, offered him affection and esteem from both officers and comrades, supplied him with a sense of power and adequately regulated his relations with authority, the element of self-concern in battle, which would lead to disruption of the effective functioning of his primary group was minimized.

- Morris Janowitz

✧✧✧

Primary groups can be highly cohesive and yet impede the goals of military organizations. Cohesive primary groups contribute to organizational effectiveness only when the standards of behaviour they enforce are articulated with the requirements of formal authority.

- Morris Janowitz

✧✧✧

We have the enemy surrounded. We are dug in and have overwhelming numbers. But enemy airpower is mauling us badly. We will have to withdraw.

- Japanese infantry officer, situation report to headquarters, Burma, WW II

✧✧✧

The tree of Liberty must be refreshed from time to time with the blood of patriots.

- Thomas Jefferson, 1787

✧✧✧

My views and feelings (are) in favour of the abolition of war—and I hope it is practicable, by improving the mind and morals of society, to lessen the disposition to war; but of its abolition I despair.

- Thomas Jefferson (1743-1826)

✧✧✧

For a people who are free, and who mean to remain so, a well-organized and armed militia is their best security.

- Thomas Jefferson: Message to Congress, November 1808

✧✧✧

From time to time, the tree of liberty must be watered with the blood of tyrants and patriots.

- Thomas Jefferson

✧✧✧

Force is the vital principle and immediate parent of despotism.

- Thomas Jefferson

✧✧✧

War is an instrument entirely inefficient toward redressing wrong; and multiplies, instead of indemnifying losses.

- Thomas Jefferson

✧✧✧

As new discoveries are made, new truths disclosed, and manners and opinions change with the change of circumstances, institutions must advance also, and keep pace with the times.

- Thomas Jefferson

✧✧✧

The price of freedom is eternal vigilance.

- Thomas Jefferson

✧✧✧

In matters of style, swim with the current;
In matters of principle, stand like a rock.

- Thomas Jefferson

✧✧✧

Every citizen should be a soldier. This was the case with the Greeks and Romans, and must be that of every free state.

-Thomas Jefferson

✧✧✧

In God's name! Let us go bravely!

- Joan of Arc, 1429

✧✧✧

Mankind has destiny beyond the grave.

- Pope John XXIII

✧✧✧

We supplicate all rulers not to remain deaf to the cry of mankind. Let them do everything in their power to save peace. By so doing they will spare the world the horrors of a war that would have disastrous consequences, such as nobody can foresee.

- Pope John XXIII

✧✧✧

Space in which to manoeuvre in the air, unlike fighting on land or sea, is practically unlimited, and . . . any number of airplanes operating defensively would seldom stop a determined enemy from getting through. Therefore the airplane was, and is, essentially an instrument of attack, not defence.

- Air Vice-Marshal J. E. 'Johnnie' Johnson, RAF.

✧✧✧

It is not possible to seal an air space hermetically by defensive tactics.

- Air Vice-Marshal J. E. 'Johnnie' Johnson, RAF.

✧✧✧

Good airplanes are more important than superiority in numbers.

- Air Vice-Marshal J. E. 'Johnnie' Johnson, RAF.

✧✧✧

Bombing is often called 'strategic' when we hit the enemy, and 'tactical' when he hits us, and is often difficult to know where one finishes and the other begins.

- Air Vice-Marshal J. E. 'Johnnie' Johnson, RAF.

✧✧✧

We carried out many trials to try to find the answer to the fast, low-level intruder, but there is no adequate defence.

- Air Vice-Marshal J. E. 'Johnnie' Johnson, RAF.

✧✧✧

There is something more important than any ultimate weapon. That is the ultimate position—the position of total control over Earth that lies somewhere out in space. That is . . . the distant future, though not so distant as we may have thought. Whoever gains that ultimate position gains control, total control, over the Earth, for the purposes of tyranny or for the service of freedom.

- Senator Lyndon B. Johnson, 1958.

✧✧✧

We are not about to send American boys nine or ten thousand miles away from home to do what Asian boys ought to be doing for themselves.

- Lyndon Johnson, October 1964

✧✧✧

Our purpose in Vietnam is to prevent the success of aggression. It is not conquest, it is not empire, it is not foreign bases, it is not domination. It is, simply put, just to prevent the forceful conquest of South Vietnam by North Vietnam.

- Lyndon Johnson

✧✧✧

Peace is a journey of a thousand miles and it must be taken one step at a time.

- Lyndon Johnson

✧✧✧

The Air Force comes in every morning and says, "Bomb, bomb, bomb." And then the State Department comes in and says, "Not now, or not there, or too much, or not at all."

- President Lyndon B. Johnson

✧✧✧

Every man thinks meanly of himself for not having been a soldier.

- Samuel Johnson

✧✧✧

Patriotism is the last refuge of a scoundrel.

- Samuel Johnson (1709-1784)

✧✧✧

As the excited passions of hostile people are of themselves a powerful enemy, both the general and his government should use their best efforts to allay them.

- Lieutenant General Antoine-Henri Baron de Jomini, 1838

✧✧✧

Woe to the general who trusts in the modern inventions, and neglects the principles of strategy.

- Lieutenant General Antoine-Henri Baron de Jomini, 1838

✧✧✧

The art of war, as generally considered, consists of five purely military branches-viz.: Strategy, Grand Tactics, Logistics, Engineering, and Tactics. A sixth and essential branch, hitherto unrecognized, might be termed Diplomacy in its relation to War. Although this branch is more naturally and intimately connected with the profession of a statesman than with that of a soldier, it cannot be denied that, if it be useless to a subordinate general, it is indispensable to every general commanding an army: it enters into all the combinations which may lead to a war, and has a connection with the various operations to be undertaken in this war; and, in this view, it should have a place in a work like this.

To recapitulate, the art of war consists of six distinct parts:

1. Statesmanship in its relation to war.
2. Strategy, or the art of properly directing masses upon the theatre of war, either for defence or for invasion.
3. Grand Tactics.
4. Logistics, or the art of moving armies.
5. Engineering,-the attack and defence of fortifications.
6. Minor Tactics.

- General Antoine-Henri Baron de Jomini

✧✧✧

War is nothing but a duel on a larger scale.

- General Antoine-Henri Baron de Jomini

✧✧✧

As the excited passions of hostile people are of themselves a powerful enemy, both the general and his government should use their best efforts to allay them.

- Lieutenant General Antoine-Henri Baron de Jomini, 1838

✧✧✧

If it seems absurd ... to mark out upon the ground an order of battle in such regular lines as would be used in tracing them on a sketch, a skilful general may nevertheless bear in mind the orders which have been indicated, and may so combine his troops on the battlefield that the arrangement shall be similar to one of them.

- General Antoine-Henri Baron de Jomini

✧✧✧

Without courageous families there would be no courageous Airmen.

- General David C. Jones

✧✧✧

I have not yet begun to fight!

- Captain John Paul Jones

During the battle on 23 September 1779 some of Jones's men cried for surrender. Captain Richard Pearson of Serapis asked Jones if he had surrendered. Jones uttered the immortal words

✧✧✧

I wish to have no connection with any ship that does not sail fast for I intend to go in harm's way.

- Captain John Paul Jones, 16 November 1778

✧✧✧

It is by no means enough that an officer be capable...He should be a gentleman of liberal education, refined, manners, punctilious courtesy, and the nicest sense of personal honor... No meritorious act of a subordinate should escape his attention, even if the reward be only one word of approval. Conversely, he should not be blind to a single fault in any subordinate.

- Captain John Paul Jones - 1775

✧✧✧

Their drills were like bloodless battles; their battles were like bloody drills.

- Flavius Josephus (37–100 A.D.), Writing of the ancient Roman army

✧✧✧

All of us have mortal bodies, composed of perishable matter, but the soul lives forever: it is a portion of the Deity housed in our bodies.

– Flavius Josephus

✧✧✧

The field of combat was a long, narrow, green-baize covered table. The weapons were words.

- Admiral C. Turner Joy (1895–1956)

✧✧✧

K

Gagan damama bajiyo paryo nishane ghao
khet jo mandyo surma ab jujhan ko dhao,
sura so pehchainye jo lare din ke het
purja purja kat mare kabhun na chadde khet.

The battle-drum (of God) beats in the sky
And lo, the target is pierced through
The gods have descended upon the battlefield
Now is the time to strike
The hero is he who fights for the oppressed
And though cut to pieces
He abandons not the field of battle

- Kabir

✧✧✧

A surprising number of government committees will make important decisions on fundamental matters with less attention than each individual would give to buying a suit.

- Herman Kahn

✧✧✧

Nuclear war is such an emotional subject that many people see the weapons themselves as the common enemy of humanity.

- Herman Kahn

✧✧✧

Authority is not power; that's coercion. Authority is not knowledge; that's persuasion, or seduction. Authority is simply that the author has the right to make a statement and to be heard.

- Herman Kahn

✧✧✧

Because of new technologies, new wealth, new conditions of domestic life and of international relations, unprecedented criteria and issues are coming up for national decision.

- Herman Kahn

✧✧✧

In a world, which is armed to its teeth with nuclear weapons, every quarrel or difference of opinion may lead to violence of a kind quite different from what is possible today.

- Herman Kahn

✧✧✧

It is immoral from almost any point of view to refuse to defend yourself and others from very grave and terrible threats, even as there are limits to the means that can be used in such defence.

- Herman Kahn

✧✧✧

World War I broke out largely because of an arms race, and World War II because of the lack of an arms race.

- Herman Kahn

✧✧✧

On the subject of diplomacy: The sword has to do the best for it does not jest.

- Karl (Charles) XII (1682–1718)
Swedish warrior king

✧✧✧

This shall hence become my music.

- Karl XII
upon hearing the sound of gunfire during his first real battle

✧✧✧

I'll likely be married to the soldier mob for better and for worse to live and to die.

- Karl XII

✧✧✧

If we cannot secure our needs for survival on the basis of law and justice, then we must be ready to secure them with army in our hands.

- Mihaly Karolyi (1918–1919), Hungarian revolutionary leader,

✧✧✧

One aspect of ... mixing of units [when casualty rates and available replacements did not align for each regiment] was that the Regimental System started to defeat its own ends. Owing to the closeness of their ties with their own regiments and their conviction that that was the best regiment in the Army, men who were suddenly drafted to another which they had been taught to believe was less good, automatically suffered a loss of morale and, in some notable cases, even refused to fight and deserted to join a battalion of their own regiment which was somewhere in the same theatre. This was a deplorable state of affairs, and happily did not occur very frequently, but it must be realized that the mixing of regiments in this way was unavoidable and was the direct result of the Regimental System as it stood.

- Lieut.-Colonel R.J.A. Kaulback, D.S.O., p.s.c., "The Regiment"

✧✧✧

While the Pakistan Army is alert to and fighting the threat posed by militancy, it remains an "India-centric" institution and that reality will not change in any significant way until the Kashmir issue and water disputes are resolved.

- General Ashfaq Parvez Kayani
Chief of Army Staff, Pakistan Army

✧✧✧

When men of equal worth fight on unequal terms, the side with the better weapons wins.

- John Keegan

✧✧✧

Politics must continue; war cannot. That is not to say that the role of the warrior is over. The world community needs, more than it has ever done, skilful and disciplined warriors who are ready to put themselves at the service of its authority. Such warriors must be properly seen as the protectors of civilisation, not its enemies.

- J. Keegan
A History of Warfare

✧✧✧

Even a pacifist should admire the military virtues.

- John Keegan

✧✧✧

I can't visualize the situation in which we nuke ourselves into extinction.

- John Keegan

✧✧✧

It's a necessary quality of a diplomat or a politician that he will compromise. Uncompromising politicians or diplomats get you into the most terrible trouble.

- John Keegan

✧✧✧

Soldiers, when committed to a task, can't compromise. It's unrelenting devotion to the standards of duty and courage, absolute loyalty to others, not letting the task go until it's been done.

- John Keegan

✧✧✧

The great Chinese classics have always said that it's better not to fight; that the clever man achieves his ends without violence; that a battle delayed is better than a battle fought.

- John Keegan

✧✧✧

The leader of men in warfare can show himself to his followers only through a mask, a mask that he must make for himself, but a mask made in such form as will mark him to men of his time and place as the leader they want and need.

- John Keegan, The Mask of Command

✧✧✧

For our purpose, however, what the soldiers did or did not read is irrelevant. For, if soldiers did not learn to fight their battles from reading books, neither is it likely that military historians learned to write their books from watching battles. Battles are extremely confusing; and confronted with the need to make sense of something he does not understand, even the cleverest, indeed pre-eminently the cleverest man, realizing his need for a language and metaphor he does not possess, will turn to look at what someone else has already made of a similar set of events as a guide for his own pen.

- John Keegan

✧✧✧

As contact with the enemy draws nearer, anticipation sharpens into fear. Its physical effects are striking. The heart beats rapidly, the face shines with sweat and the mouth grows dry—so dry that men often emerge from battle with blackened mouths and chapped lips.

- John Keegan

✧✧✧

War, in Christian theology, is a sinful activity, unless carried on within a framework of rules which few commanders are in practice able to obey; in particular those which demand that he shall have a just aim and a reasonable expectation of victory. Any objective study quickly reveals, however, that most wars are begun for reasons which have nothing to do with justice, have results quite different from those proclaimed as their objects, if indeed they have any clear-cut result at all, and visit during their course a great deal of casual suffering on the innocent.

- John Keegan, The Face of Battle (1976)

✧✧✧

Behind the Edinburghs, the four Tyneside Irish battalions of the 103rd Brigade underwent a bizarre and pointless massacre. ... This decision gave the last brigade a mile of open ground to cover before it reached its own front line, a safe enough passage if the enemy's machine-guns had been extinguished, otherwise a funeral march. A sergeant of the 3rd Tyneside Irish (26th Northumberland Fusiliers) describes how it was: "I could see, away to my left and right, long lines of men. Then I heard the 'patter, patter' of machine-guns in the distance. By the time I'd gone another ten yards there seemed to be only a few men left around me; by the time I had gone twenty yards, I seemed to be on my own. Then I was hit myself." Not all went down so soon. A few heroic souls pressed on to the British front line, crossed no-man's land and entered the German trenches. But the brigade was destroyed; one of its battalions had lost over 600 men killed or wounded, another, 500; the brigadier and two battalion commands had been hit, a third lay dead. Militarily, the advance had achieved nothing. Most of the bodies lay on territory British before the battle had begun.

- John Keegan, The Face of Battle (1976)

✧✧✧

The "tenacity and resolution" of a commander in the face of great danger is a mightily inspiring thing to a common soldier.

- John Keegan and Holmes
Soldiers: A History of Men in Battle

✧✧✧

A Roman centurion named Petronius Arbiter said in 210 B.C. "We trained hard, but it seemed that every time we were beginning to form up into teams, we would be reorganized. I was to learn later in life that we tend to meet any new situation by reorganizing: and a wonderful method it can be for creating the illusion of progress while producing confusion, inefficiency, and demoralization."

- Anthony Kellett
Combat Motivation: The Behaviour of Soldiers in Battle

✧✧✧

A war regarded as inevitable or even probable, and therefore much prepared for, has a good chance of eventually being fought.

- George F. Keenan, The Cloud of Danger

✧✧✧

A young man who does not have what it takes to perform military service is not likely to have what it takes to make a living.

- John F. Kennedy (1917–1963)
35th President of the United States

✧✧✧

Ask not what your country can do for you; ask what you can do for your country.

- John F. Kennedy

✧✧✧

Mankind must put an end to war, or war will put an end to mankind. War will exist until that distant day when the conscientious objector enjoys the same reputation and prestige that the warrior does today.

- John F. Kennedy
Speech to UN General Assembly, 25 September 1961

✧✧✧

Let every nation know, whether it wishes us well or ill, that we shall pay any price, bear any burden, meet any hardship, support any friend, oppose any foe, to assure the survival and success of liberty.

- John F. Kennedy

✧✧✧

Change is the law of life. And those who look only to the past or the present are certain to miss the future.

- John F. Kennedy, Speech at Frankfurt, 25 June 1963

✧✧✧

A man may die, nations may rise and fall, but an idea lives on.

- John F. Kennedy

✧✧✧

Forgive your enemies, but never forget their names.

- John F. Kennedy

✧✧✧

Events of October 1962 indicated, as they had all through history, that control of the sea means security. Control of the seas can mean peace. Control of the seas can mean victory. The United States must control the seas if it is to protect your security....

- President John F. Kennedy
6 June 1963, on board USS Kitty Hawk

✧✧✧

I can imagine no more rewarding a career. And any man who may be asked in this century what he did to make his life worthwhile, I think can respond with a good deal of pride and satisfaction: 'I served in the United States Navy.

- President John F. Kennedy
1 August 1963, at the U. S. Naval Academy

✧✧✧

Let us never negotiate out of fear. But let us never fear to negotiate.

- John F. Kennedy

✧✧✧

Only those who dare to fail greatly can ever achieve greatly.

- Senator Robert F. Kennedy

✧✧✧

It is from numberless diverse acts of courage and belief that human history is shaped. Each time a person stands up for an ideal, or acts to improve the lot of others, or strikes out against injustice, he sends forth a tiny ripple of hope. That ripple builds others. Those ripples—crossing each other from a million different centres of energy—build a current that can sweep down the mightiest walls of oppression and injustice.

- Senator Robert F. Kennedy

✧✧✧

Economic prosperity does not always translate into military effectiveness: still, all the major power shifts in the world's military power balances have followed alterations in the productive balances.

- Paul Kennedy

✧✧✧

All the major shifts in the world's military-power balances have followed alterations in the productive balances. Victory has always gone to the side with the greatest material resources.

- Paul Kennedy

✧✧✧

They happened to be at the right place at the right time.

- Paul Kennedy

✧✧✧

The triumph of any one Great Power ...., or the collapse of another, has usually been the consequence of lengthy fighting by its armed forces; but it has also been the consequences of the more or less efficient utilization of the state's productive economic resources in wartime, and, further in the background, of the way in which that state's economy had been rising or falling, relative to the other leading nations, in the decades preceding the actual conflict. For that reason, how a Great Power's position steadily alters in peacetime is as important as how it fights in wartime.

- Paul Kennedy

✧✧✧

Air power is like poker. A second-best hand is like none at all—it will cost you dough and win you nothing.

- General George Kenney,
Commander of Allied Air Forces in the Southwest Pacific, 1942-45.

✧✧✧

History is not kind to nations that go to sleep. Pearl Harbour woke us up and we managed to win, although we are already forgetting the dark days when victory was uncertain, when it looked as though the scales might be tipped the other way.

- General George C. Kenney

✧✧✧

Everything, everything in war is barbaric....But the worst barbarity of war is that it forces men collectively to commit acts against which individually they would revolt with their whole being.

- Ellen Key, War, Peace and the Future

✧✧✧

Every leader, and every regime, and every movement, and every organization that steps across the line to terrorism must be banished from the discourse of civilized human life.

- Alan Keyes, 21 April 2002

✧✧✧

In all operations, a moment arrives when brave decisions have to be made if an enterprise is to be carried through.

- Sir Roger Keyes

✧✧✧

The inevitable never happens. It is the unexpected, always.

- Lord Keynes

✧✧✧

Alike for those who today prepare
And those who for a tomorrow stare
A Muezzin from the Tower of Darkness cries
Fools! Your reward is neither here nor there.

- Omar Khayyam

✧✧✧

The Moving Finger writes; and, having writ,
Moves on: nor all thy Piety nor Wit
Shall lure it back to cancel half a Line,
Nor all thy Tears wash out a Word of it.

- Omar Khayyam

✧✧✧

At least my Gun is OK!!!
Its firing, and I shall continue to fire as long as it works.

- 2 Lieut Arun Khetrapal , Param Vir Chakra (17 Poona Horse)

His tank was engulfed in flames while facing enemy fire. His tank was the only working one among three others. He said these words to his commander when ordered to evacuate the vehicle. He died soon after destroying the remaining Pakistani tanks—his tank burst into flames.

✧✧✧

The guiding principle of the strategy for our whole resistance must be to prolong the war. To protract the war is the key to victory. Why must the war be protracted? ... If we throw the whole of our forces into a few battles to try to decide the outcome, we shall certainly be defeated and the enemy will win. On the other hand, if while fighting we maintain our forces, expand them, train our army and people, learn military tactics ... and at the same time wear down the enemy forces, we shall weary and discourage them in such a way that, strong as they are, they will become weak and will meet defeat instead of victory.

- Dang Xuan Khu, second in command to Ho Chi Minh, Primer for Revolt

✧✧✧

Life has to be lived forward, but it has to be understood backward.

- Soren Kierkegaard

✧✧✧

It is so hard to believe because it is so hard to obey.

- Soren Kierkegaard

✧✧✧

One can advise comfortably from a safe port.

- Soren Kierkegaard

✧✧✧

During the first period of a man's life the greatest danger is not to take the risk.

- Soren Kierkegaard

✧✧✧

How could they possibly be Japanese planes?

- Admiral Husband E. Kimmel (1882–1968)
Commander-in-chief, U.S. Pacific Fleet at the time of the Japanese attack on Pearl Harbour

✧✧✧

Hold what you've got and hit them where you can.

- Admiral of the Fleet Ernest J. King, December 1941

✧✧✧

If a man hasn't discovered something that he will die for, he isn't fit to live.

- Martin Luther King

✧✧✧

IF

If you can keep your head when all about you
Are losing theirs and blaming it on you,
If you can trust yourself when all men doubt you,
But make allowance for their doubting too;
If you can wait and not be tired by waiting,
Or being lied about, don't deal in lies,
Or being hated, don't give way to hating,
And yet don't look too good, nor talk too wise:
If you can dream—and not make dreams your master;
If you can think—and not make thoughts your aim;
If you can meet with Triumph and Disaster
And treat those two impostors just the same;
If you can bear to hear the truth you've spoken
Twisted by knaves to make a trap for fools,
Or watch the things you gave your life to, broken,
And stoop and build 'em up with worn-out tools:

If you can make one heap of all your winnings
And risk it on one turn of pitch-and-toss,
And lose, and start again at your beginnings
And never breathe a word about your loss;
If you can force your heart and nerve and sinew
To serve your turn long after they are gone,
And so hold on when there is nothing in you
Except the Will which says to them: 'Hold on!'

If you can talk with crowds and keep your virtue,
'Or walk with Kings—nor lose the common touch,
if neither foes nor loving friends can hurt you,
If all men count with you, but none too much;
If you can fill the unforgiving minute
With sixty seconds' worth of distance run,
Yours is the Earth and everything that's in it,
And—which is more—you'll be a Man, my son!

- Rudyard Kipling

✧✧✧

The 'eathen in 'is blindness must end where 'e began. But the backbone of the Army is the non-commissioned man!

-Rudyard Kipling

✧✧✧

We have a thousand reasons for failure but not a single excuse.

- Rudyard Kipling

✧✧✧

Give me to live and love in the old, bold fashion;
A soldier's billet at night and a soldier's ration;
A heart that leaps to the fight with a soldier's passion.
For I hold as a simple faith there's no denying:
The trade of a soldier's the only trade worth plying;
The death of a soldier's the only death worth dying.
... I'll die as a soldier dies on the Field of Glory.
... Death in my boots may-be, but fighting, fighting.

- Rudyard Kipling,
The Song of the Soldier born
Rhymes of a Red Cross Man, 1916

✧✧✧

It's Tommy this, and Tommy that,
And chuck him out the brute,
But it's 'Saviour of his Country,'
When the guns begin to shoot!
For God and the soldier we adore, In time of danger, not before!
The danger passed, and all things righted,
God is forgotten and the soldier slighted.

- Rudyard Kipling

✧✧✧

When you're wounded out on Afghanistan's plains
And the women come out to cut up what remains,
Then just roll to your rifle and blow out your brains
And die like a good British soldier!

- Rudyard Kipling

✧✧✧

If your officer's dead and the sergeants look white,
Remember it's ruin to run from a fight:
So take open order, lie down, and sit tight,
And wait for supports like a soldier.

- Rudyard Kipling,
The Young British Soldier

✧✧✧

Give me to live and love in the old, bold fashion;
A soldier's billet at night and a soldier's ration;
A heart that leaps to the fight with a soldier's passion.
For I hold as a simple faith there's no denying:
The trade of a soldier's the only trade worth plying;
The death of a soldier's the only death worth dying.
... I'll die as a soldier dies on the Field of Glory.
... Death in my boots may-be, but fighting, fighting.

- Rudyard Kipling
The Song of the Soldier-born, Rhymes of a Red Cross Man, 1916

✧✧✧

. . . 't isn't the best drill, though drill is nearly everything, that hauls a Regiment through Hell and out on the other side. It's the man who knows how to handle men—goat-men, swine-men, dog-men, and so on.

- Rudyard Kipling from "Only a Subaltern,"
The Man Who Would be King and other stories

✧✧✧

No proposition Euclid wrote
No formula the text books know
Will turn the bullet from your coat
Or ward the tulwar's downward blow

- Rudyard Kipling

✧✧✧

Odd things always go wrong in the preparation for a battle.

- Kippenberger, Infantry Brigadier

✧✧✧

War, if reason prevails, is waged to obtain a better peace than that which existed prior to the hostilities.

- Bela K. Kiraly
Hungarian general and American military historian

✧✧✧

We have war when at least one of the parties to a conflict wants something more than it wants peace.

- Jeane J. Kirkpatrick

✧✧✧

Straying off course is not recognized as a capital crime by civilized nations.

- Jeane Kirkpatrick, on the Soviet destruction of Korean Airways Flight 007

✧✧✧

Foreign policy must begin with some definition of what constitutes a vital interest—a change in the international environment so likely to undermine the national security that it must be resisted no matter what form the threat takes or how ostensibly legitimate it appears.

- Henry Kissinger

✧✧✧

Never before has a new world order had to be assembled from so many different perceptions, or on so global a scale. Nor has any previous order had to combine the attributes of the historic balance-of-power systems with global democratic opinion and the exploding technology of the contemporary period.

- Henry Kissinger
Diplomacy

✧✧✧

Paradoxically, the generality of this dissatisfaction is a condition of stability, because were any one power totally satisfied, all others would have to be totally dissatisfied and a revolutionary situation would ensue. The foundation of a stable order is the relative security—and therefore the relative insecurity—of its members.

- Henry Kissinger

✧✧✧

There are only two roads to stability: one is hegemony and the other is equilibrium.

- Henry Kissinger

✧✧✧

A leader does not deserve the name unless he is willing occasionally to stand alone.

- Henry Kissinger

✧✧✧

A leader who confines his role to his people's experience dooms himself to stagnation; a leader who outstrips his people's experience runs the risk of not being understood.

- Henry Kissinger

✧✧✧

High office teaches decision making, not substance. It consumes intellectual capital; it does not create it. Most high officials leave office with the perceptions and insights with which they entered; they learn how to make decisions but not what decisions to make.

- Henry Kissinger

✧✧✧

If it's going to come out eventually, better have it come out immediately.

- Henry Kissinger

✧✧✧

If you don't know where you are going, every road will get you nowhere.

- Henry Kissinger

✧✧✧

"The task of the leader is to get his people from where they are to where they have not been.

- Henry Kissinger

✧✧✧

Ninety percent of the politicians give the other ten percent a bad reputation.

- Henry Kissinger

✧✧✧

The superpowers often behave like two heavily armed blind men feeling their way around a room, each believing himself in mortal peril from the other, whom he assumes to have perfect vision.

- Henry Kissinger

✧✧✧

To be absolutely certain about something, one must know everything or nothing about it.

- Henry Kissinger

✧✧✧

No foreign policy—no matter how ingenious—has any chance of success if it is born in the minds of a few and carried in the hearts of none.

- Henry Kissinger

✧✧✧

The illegal we do immediately. The unconstitutional takes a little longer.

- Henry Kissinger
New York Times, 28 October 1973

✧✧✧

The conventional army loses if it does not win. The guerrilla wins if he does not lose.

- Henry Kissinger

✧✧✧

Power is the ultimate aphrodisiac.

- Henry Kissinger

✧✧✧

Don't talk to me about atrocities in war; all war is an atrocity.

- Lord Kitchener

✧✧✧

Low intensity conflict is a politico-military confrontation between contending states or groups, below the level of conventional war and above the routine peaceful competition among states. It involves protracted struggle of competing principles and ideologies, and ranges from subversion to the use of armed forces. It is waged by a combination of means, employing political, economic, informational and military instruments. Such conflicts are often local in nature but contain regional and global security implications.

- Frank Kitson (Low Intensity Operations)

✧✧✧

Remember one lesson from the Vietnam era: Those who ordered the meal were not there when the waiter brought the check.

- General William A. Knowlton
Ethics and Decision-Making

✧✧✧

The understanding of past events and problems can be part of a learning process that assists us in understanding present events and problems. Without knowing how past societies coped with problems similar to our own, the options they considered, the choices they made, and the consequences entailed by these choices, we can only arrive at a flat, poorly constructed understanding of our present problems.

- Klaus Knorr
Historical Dimensions of National Security Problems, 1976

✧✧✧

Soldiers are men...most apt for all manner of services and best able to support and endure the infinite toils and continual hazards of war.

- Henry Knyvett

✧✧✧

The most persistent sound which reverberates through men's history is the beating of war drums.

- Arthur Koestler,
Janus: A Summing Up

✧✧✧

Send one plane it's a sortie; send two it's a flight; send four and it's a test of airpower.

- Richard Kohn

✧✧✧

How is the world ruled and how do wars start?
Diplomats tell lies to journalists and then believe what they read.

- Karl Kraus (1874–1936)

✧✧✧

Being ready is not what matters. What matters is winning after you get there.

- Lieutenant General V.H. Krulak,
US Marine Corps

✧✧✧

The more the two of us pull, the tighter the knot will be tied. In this process, a moment may come when that knot will be tied so tight that even he who tied it will not have the strength to untie it, and then it will be necessary to cut that knot, and what that would mean is not for me to explain to you, because you understand perfectly of what terrible forces our countries dispose. Consequently, if there is no intention to tighten that knot, and thereby doom the world to a catastrophe of thermonuclear war, then let us not only relax the forces pulling on the ends of the rope, let us take measures to untie that knot. We are ready for this.

- Nikita Krushchev
Russian Prime Minister, in a letter to President John Kennedy

✧✧✧

But let me return to the enemy breakthrough in the Kiev area, the encirclement of our group, and the destruction of the 37th Army. Later, the Fifth Army also perished ... All of this was senseless, and from the military point of view, a display of ignorance, incompetence, and illiteracy. ... There you have the result of not taking a step backward. We were unable to save these troops because we didn't withdraw them, and as a result we simply lost them. ... And yet it was possible to allow this not to happen.

- Nikita Krushchev

✧✧✧

Bombs do not choose. They will hit everything.

- Nikita Krushchev

✧✧✧

In a fight you don't stop to choose your cudgels.

- Nikita Krushchev

✧✧✧

Politicians are the same all over. They promise to build bridges even when there are no rivers.

- Nikita Krushchev

✧✧✧

Whether in an advantageous position or a disadvantageous one, the opposite state should be always present to your mind.

- Ts'ao Kung

✧✧✧

Bravery without forethought causes a man to fight blindly and desperately like a mad bull. Such an opponent must not be encountered with brute force, but may be lured into an ambush and slain

- Ts'ao Kung

✧✧✧

L

If inciting people to do that [9/11] is terrorism, and if killing those who kill our sons is terrorism, then let history be witness that we are terrorists.

- Osama Bin Laden, interview, October 2001

✧✧✧

India ... was a hard country for toughened veterans; it could be hell for a boy soldier. One of them, serving in the 71st Regiment—the Highland Light Infantry—wrote: "This was the first blood I had ever seen shed in battle; the first time the cannon had roared in my hearing charged with death. I was not yet seventeen years of age, and had not been six months from home. My limbs bending under me with fatigue, in a sultry clime, the musket and accoutrements that I was forced to carry were insupportably oppressive. Still, I bore all with invincible patience. During the action, the thought of death never crossed my mind. After the firing commenced, a still sensation stole over my whole frame, a firm determined torpor, bordering on insensibility. I heard an old soldier answer, to a youth like myself, who asked what he should do during the battle, 'Do your duty.'

- John Laffin, Boys in Battle, 1966

✧✧✧

If you're in a fair fight, you didn't plan it properly.

- Nick Lappos, Chief R&D Pilot, Sikorsky Aircraft.

✧✧✧

When the rules of the game prove unsuitable for victory, the gentlemen of England change the rules.

- Harold Laski

✧✧✧

I must follow them. I am their leader.

- Andrew Bonar Law, 1858-1923

✧✧✧

With two thousand years of examples behind us we have no excuse, when fighting, for not fighting well.

- T.E. Lawrence (of Arabia) (1888–1935)

✧✧✧

Nine-tenths of tactics are certain and taught in books: but the irrational tenth is like the kingfisher flashing across the pond and that is the test of generals. It can only be ensured by instinct, sharpened by thought practicing the stroke so often at the crisis it is as natural as a reflex.

- T.E. Lawrence
The Science of Guerrilla Warfare

✧✧✧

They taught me that no man could be their leader except he ate the ranks' food, wore their clothes, lived level with them, and yet appeared better in himself.

- T.E. Lawrence
The Seven Pillars of Wisdom

✧✧✧

All men dream; but not equally. Those who dream by night in the dusty recesses of their minds wake in the day to find that it was vanity; but the dreamers of the day are dangerous men, for they may act out their dreams with open eyes, to make it possible.

- T. E. Lawrence

✧✧✧

An opinion can be argued with; a conviction is best shot.

- T.E. Lawrence

✧✧✧

Do not try to do too much with your own hands. Better the Arabs do it tolerably than you do it perfectly. It is their war, and you are to help them, not to win it for them.

- T.E. Lawrence
Article 15 of "27 Articles", published in the Arab Bulletin of 20 August 1917, as a guide for British officers fighting alongside Arab tribesmen

✧✧✧

Don't give up the ship!

- Captain James Lawrence
Tradition has it that he said these heroic words after being mortally wounded in the engagement between his ship, the U.S. frigate Chesapeake, and HMS Shannon on 1 June 1813.
Although Chesapeake was forced to surrender, Captain Lawrence's words lived on as a rallying cry during the war. Oliver Hazard Perry honoured his dead friend Lawrence when he had the motto sewn onto the private battle flag flown during the Battle of Lake Erie, 10 September 1813

✧✧✧

However incomprehensible the acts of the terrorists appear to be, our judges, our policemen, and our politicians must never be allowed to forget that terrorism is an activity of the fellow human beings.

- Edmond Leach

✧✧✧

Duty is the sublimest word in our language. Do your duty in all things. You cannot do more. You should never do less.

- Robert E. Lee
US Confederate General

✧✧✧

It is well that war is so terrible, else we should grow too fond of it.

- Robert E. Lee
US Confederate General

✧✧✧

We made a great mistake in the beginning of our struggle, and I fear, in spite of all we can do, it will prove to be a fatal mistake. We appointed all our worst generals to command our armies, and all our best generals to edit the newspapers.

- Robert E. Lee

✧✧✧

The fixing of bayonets is more than a fixing of steel to a rifle it puts iron into the soul of the soldier doing the fixing.

- Robert E. Lee
US Confederate General

✧✧✧

I think and work with all my power to bring the troops to the right place at the right time; then I have done my duty. As soon as I order them forward into battle, I leave my army in the hands of God.

- Robert E. Lee, US Confederate General

✧✧✧

1. Legionnaire: you are a volunteer serving France faithfully and with honour.
2. Every Legionnaire is your brother-at-arms, irrespective of his nationality, race or creed. You will demonstrate this by an unwavering and straight forward solidarity which must always bind together members of the same family.
3. Respectful of the Legion's traditions, honouring your superiors, discipline and comradeship are your strength, courage and loyalty your virtues.
4. Proud of your status as a legionnaire, you will display this pride, by your turnout, always impeccable, your behaviour, ever worthy, though modest, your living-quarters, always tidy.
5. An elite soldier: you will train vigorously, you will maintain your weapons as if it were your most precious possession, you will keep your body in the peak of condition, always fit.
6. A mission once given to you becomes sacred to you, you will accomplish it to the end and at all costs.
7. In combat: you will act without relish of your tasks, or hatred; you will respect the vanquished enemy and will never abandon neither your wounded nor your dead, nor will you under any circumstances surrender your arms.

- French Foreign Legion (Code of Honour)

✧✧✧

I don't mind being called tough, since I find in this racket it's the tough guys who lead the survivors.

- General Curtis LeMay, US Air Force

✧✧✧

Tell the Vietnamese they've got to draw in their horns or we're going to bomb them back into the Stone Age.

- General Curtis LeMay, May 1964

✧✧✧

If we maintain our faith in God, love of freedom, and superior global air power, the future [of the US] looks good.

- General Curtis LeMay

✧✧✧

Saving the lives of your fellow airmen is the most extraordinary kind of heroism that I know.

- General Curtis E. LeMay

✧✧✧

The soundest strategy in war is to postpone operations until the moral disintegration of the enemy renders the delivery of the mortal blow both possible and easy.

- V. I. Lenin,. Russian revolutionary leader

✧✧✧

A lie told often enough becomes the truth.

- Vladimir Ilyich Lenin

✧✧✧

Quantity has a quality all its own.

- Vladimir Ilyich Lenin

✧✧✧

Insurrection is an art as much as war.

- Vladimir Ilyich Lenin

✧✧✧

Pose the enemy a dilemma, not a problem.

- Lt Col Robert R. Leonhard, U.S. Army

✧✧✧

Armies don't innovate; people innovate. The single most important quality in a professional military officer is the ability to innovate. In an age of increased technological advancement and sociopolitical upheaval, the number of transitions in military art and science will multiply, with the result that in the span of an officer's career warfare will undergo several dramatic changes. The failure of the officer corps to keep up with change can result in national disaster. But how can a military establishment-by its very nature a conservative establishment-systematically and consistently train its officers to innovate?

- Lt Col Robert R. Leonhard, U.S. Army
Fighting by Minutes: Time and the Art of War

✧✧✧

In manoeuvre warfare, we attempt not to destroy the entire enemy force but to render most of it irrelevant.

- Lt Col Robert R. Leonhard, U.S. Army

✧✧✧

Courage is not merely one of the virtues but the form of every virtue at the testing point, which means at the point of highest reality.

- C. S. Lewis

✧✧✧

The navy of any great power . . . has the dream to have one or more aircraft carriers. The question is not whether you have an aircraft carrier, but what you do with your aircraft carrier.

- Major General Qian Lihua, director of the Chinese Defence Ministry's Foreign Affairs Office, on China wanting to start carrier operations. 'Financial Times,' 16 November 2008.

✧✧✧

I am not bound to win, but I am bound to be true. I am not bound to succeed, but I am bound to live by the light that I have. I must stand with anybody that stands right, stand with him while he is right, and part with him when he goes wrong.

- Abraham Lincoln (1809–1865)
16th President of the United States

✧✧✧

Peace will come soon to stay, and so come as to be worth keeping in all future time. It will then have proved that among free men there can be no successful appeal from the ballot to the bullet, and that they who take such appeal are sure their cases and pay the costs.

- Abraham Lincoln

✧✧✧

With malice toward none, with charity for all, with firmness in the right, as God gives us to see the right, let us strive on to finish the world we are in, to bind up the nation's wounds.

- Abraham Lincoln: Second Inaugural Address, 1865.

✧✧✧

The shepherd drives the wolf from the sheep's for which the sheep thanks the shepherd as his liberator, while the wolf denounces him for the same act as the destroyer of liberty. Plainly, the sheep and the wolf are not agreed upon a definition of liberty.

- Abraham Lincoln

✧✧✧

We shall meanly lose or nobly save the last hope of earth.

- Abraham Lincoln

✧✧✧

Soldiers—You are about to return to your homes and your friends, after having, as I learn, performed in camp a comparatively short term of duty in this great contest. I am greatly obliged to you, and to all who have come forward at the call of their country. I wish it might be more generally and universally understood what the country is now engaged in. We have, as all will agree, a free Government, where every man has a right to be equal with every other man. In this great struggle, this form of Government and every form of human right is endangered if our enemies succeed. There is more involved in this contest than is realized by every one. There is involved in this struggle the question whether your children and my children shall enjoy the privileges we have enjoyed. I say this in order to impress upon you, if you are not already so impressed, that no small matter should divert us from our great purpose. There may be some irregularities in the practical application of our system. It is fair that each man shall pay taxes in exact proportion to the value of his property; but if we should wait before collecting a tax to adjust the taxes upon each man in exact proportion with every other man, we should never collect any tax at all. There may be mistakes made sometimes; things may be done wrong while the officers of the Government do all they can to prevent mistakes. But I beg of you, as citizens of this great Republic, not to let your minds to carried off from the great work we have before us. This struggle is too large for you to be diverted from it by any small matter. When you return to your homes rise up to the height of a generation of men worthy of a free Government, and we will carry out the great work we have commenced. I return to you my sincere thanks, soldiers, for the honour you have done me this afternoon.

- Abraham Lincoln
Speech to the One Hundred Sixty-Fourth Ohio Regiment Washington, D.C., 18 August 1864

✧✧✧

The man who can't make a mistake can't make anything.

- Abraham Lincoln

✧✧✧

Public sentiment is everything. With public sentiment nothing can fail. Without it nothing can succeed. He who moulds opinion is greater than he who enacts laws.

- Abraham Lincoln

✧✧✧

Peace will come soon to stay, and so come as to be worth keeping in all future time. It will then have proved that among free men there can be no successful appeal from the ballot to the bullet, and that they who take such appeal are sure their cases and pay the costs.

- Abraham Lincoln

✧✧✧

Our cause, then, must be entrusted to, and conducted by, its own undoubted friends-those whose hands are free, whose hearts are in the work-who do care for the result. Two years ago the Republicans of the nation mustered over thirteen hundred thousand strong. We did this under the single impulse of resistance to a common danger, with every external circumstance against us. Of strange, discordant, and even, hostile elements, we gathered from the four winds, and formed and fought the battle through, under the constant hot fire of a disciplined, proud, and pampered enemy. Did we brave all then to falter now?-now when that same enemy is wavering, dissevered, and belligerent? The result is not doubtful. We shall not fail-if we stand firm, we shall not fail. Wise councils may accelerate or mistakes delay it, but, sooner or later, the victory is sure to come

- Abraham Lincoln

✧✧✧

Military glory—that attractive rainbow that rises in showers of blood.

- Abraham Lincoln

✧✧✧

If this is coffee, please bring me some tea; but if this is tea, please bring me some coffee.

- Abraham Lincoln

✧✧✧

Four score and seven years ago our fathers brought forth on this continent a new nation, conceived in liberty, and dedicated to the proposition that all men are created equal.

Now we are engaged in a great civil war, testing whether that nation, or any nation, so conceived and so dedicated, can long endure. We are met on a great battle-field of that war. We have come to dedicate a portion of that field, as a final resting place for those who here gave their lives that that nation might live. It is altogether fitting and proper that we should do this.

But, in a larger sense, we cannot dedicate, we cannot consecrate, we cannot hallow this ground. The brave men, living and dead, who struggled here, have consecrated it, far above our poor power to add or detract. The world will little note, nor long remember what we say here, but it can never forget what they did here. It is for us the living, rather, to be dedicated here to the unfinished work which they who fought here have thus far so nobly advanced. It is rather for us to be here dedicated to the great task remaining before us—that from these honoured dead we take increased devotion to that cause for which they gave the last full measure of devotion—that we here highly resolve that these dead shall not have died in vain—that this nation, under God, shall have a new birth of freedom—and that government of the people, by the people, for the people, shall not perish from the earth.

- Abraham Lincoln, Gettysburg Address, 19 November 1863

✧✧✧

He who does something at the head of one regiment will eclipse him who does nothing at the head of a hundred.

- Abraham Lincoln

✧✧✧

Let us have faith that right is might and in that faith let us to the end dare to do our duty as we understand it.

- Abraham Lincoln

✧✧✧

No man is good enough to govern another man without that other's consent.

- Abraham Lincoln

✧✧✧

The probability that we may fail in the struggle ought not to deter us from the support of a cause we believe to be just.

- Abraham Lincoln

✧✧✧

From the complaints of the soldiers, I guess that's about all any of you do pay.

- Abraham Lincoln

To an Army quartermaster who said that he would like to pay his respects

✧✧✧

Force is all-conquering, but its victories are short-lived.

- Abraham Lincoln

✧✧✧

A manoeuvre warfare military believes it is better to have high levels of initiative among subordinate officers, with a resultant rapid Boyd Cycle, even if the price is some mistakes.

- William S. Lind, Manoeuvre Warfare Handbook, 1985

✧✧✧

Critics of these faster moving forms of combat characterized by initiative at the low level fear that they will lead to groups of men moving willy-nilly about the battlefield and that commanders will lose control. This need not be so. That is why we have control measures. The boundary, the limit of advance, the phase line, can still be used. It must be remembered that these control measures should serve their function, but not be rigid lines that cannot be changed or ignored when the situation changes. They should be kept to a minimum and must always be flexible. The tactics must never follow the control measures. On the contrary, the control measures must follow the tactics.

- William S. Lind, Manoeuvre Warfare Handbook, 1985

✧✧✧

Gaps are found by delegating authority down to the lowest level, so that small unit commanders can find gaps and immediately start exploiting them without delay. If we are always going to wait for directions from above, our force is going to be very slow moving, so this concept of surfaces and gaps depends on initiative at low levels.

- William S. Lind, Manoeuvre Warfare Handbook, 1985

✧✧✧

Conflict can be seen as a time competitive observation-orientation-decision-action cycles. Each party to a conflict begins by observing. He observes himself, his physical surroundings and his enemy. On the basis of his observation, he orients, that is to say, he makes a mental image or 'snapshot' of his situation. On the basis of this orientation he makes a decision. He puts the decision into effect, i.e. he acts. Then because the action has changed the situation, he observes again, and starts the process anew. His actions follow this cycle...

- William Lind. The Manoeuvre Warfare Handbook, The OODA Loop

✧✧✧

The essence of mission tactics: "The subordinate decides what to do even if it means that the order issued by his senior now should be changed or adjusted. The mission assigned is sacred. The mission is the output that the commander wants. That does not change. But how that output is to be achieved may change, and it is up to the intelligent subordinate to decide whether or not it has."

- William S. Lind, Manoeuvre Warfare Handbook, 1985

✧✧✧

But usually your mission order will have the ability to endure time better if you explain to your subordinate why he is carrying the mission out: 'In order to—.' This gives the order the quality that Erich von Manstein called 'long-term.' It can endure the test of time. Your commander can lose communication with you yet you can still carry out his intent because you know what he wanted and you can continue to act within his intent for a long time without checking back.

- William S. Lind, Manoeuvre Warfare Handbook, 1985

✧✧✧

But I have seen the science I worshiped, and the airplane I loved, destroying the civilization I expected them to serve.

- Charles A. Lindbergh, 'Time,' 26 May 1967

✧✧✧

Generals speak often of their military duty to their superiors, but never of their duty to their soldiers.

- Helmut Lindmann, German military historian

✧✧✧

A nation has security when it does not have to sacrifice its legitimate interests to avoid war and is able, if challenged, to maintain them by war.

- Walter Lippmann

✧✧✧

Nobody has yet found a way of bombing that can prevent foot soldiers from walking.

- Walter Lippmann

✧✧✧

The final test of a leader is that he leaves behind him in other men the conviction and the will to carry on.

- Walter Lippmann, 1889-1974

✧✧✧

No art or science is as difficult as that of war.

-Henry Humphrey Evans Lloyd, 18th century British military analyst

✧✧✧

The fluttering of a butterfly's wing in Rio de Janeiro, amplified by atmospheric currents, could cause a tornado in Texas two weeks later.

- Edward Lorenz

✧✧✧

Commanders in time of peace should therefore make it their duty to encourage initiative in their subordinates, instead of checking it, as is the case too often. Subordinates should be taken to task only when their action was taken thoughtlessly—without a good reason.

- Major General Baron Hugo von Freytag-Loringhoven
The Power of Personality In War

✧✧✧

Better a thousand times to die with glory than live without honour.

- Louis VI of France, 1081–1137

✧✧✧

For a soldier, leadership is that power to inspire soldiers in battle, or unforeseen crisis, where others fail; the quality possessed by the individual, of whatever rank, whose courage lifts lesser men with the certainty that all will go well as long as he is there.

- Lord Lovat

✧✧✧

Stone walls do not a prison make,
Nor iron walls a cage;
Minds innocent and quiet take,
That for a hermitage;
If I have freedom in my love'
And in my soul am free;
Angels alone that soar above,
Enjoy such liberty.

- Richard Lovelace (1618-1657)

✧✧✧

A man who wants to lead the orchestra must turn his back on the crowd.

- Max Lucado

✧✧✧

A study of past battles have shown that men have time and again, been inspired above and beyond the call of duty by the traditions of a unit.

- James D Lunt

✧✧✧

Hard times make for soft principles.

- Gavin Lyall

✧✧✧

The safest place in Korea was right behind a platoon of Marines. Lord, how they could fight!

- Maj. Gen. Frank Lowe, US Army

✧✧✧

M

If your country's worth living in, it's worth fighting for ... you can't fight a war without losing lives. Although no one wants killing, sometimes it has to be. To keep your country free sometimes you have to fight and die. It was to be a great honour to us.

- Billy Mabin
Shankill Road, 1916 Somme victim of the 36th (Ulster) Division

✧✧✧

Then out spake brave Horatius,
The Captain of the Gate:
To every man upon this earth
Death cometh soon or late.
And how can man die better
Than facing fearful odds,
For the ashes of his fathers,
And the temples of his gods.

- Thomas Macaulay
(Lays of Ancient Rome, Horatius, Stanza XXVII)

✧✧✧

The measure of a man's real character is what he would do if he knew he would never be found out.

- Thomas Macaulay

✧✧✧

It is impossible for us, with our limited means, to attempt to educate the body of the people. We must at present do our best to form a class who may be interpreters between us and the millions whom we govern; a class of persons, Indian in blood and colour, but English in taste, in opinions, in morals, and in intellect. To that class we may leave it to refine the vernacular dialects of the country, to enrich those dialects with terms of science borrowed from the Western nomenclature, and to render them by degrees fit vehicles for conveying knowledge to the great mass of the population.

- Thomas Macaulay
Minute on Indian Education, 1835

✧✧✧

Many politicians are in the habit of laying it down as a self-evident proposition that no people ought to be free till they are fit to use their freedom. The maxim is worthy of the fool in the old story who resolved not to go into the water till he had learned to swim.

- Thomas Macaulay

✧✧✧

The knowledge of the theory of logic has no tendency whatever to make men good reasoners.

- Thomas Macaulay

✧✧✧

History is made up of the bad actions of extraordinary men and woman. All the most noted destroyers and deceivers of our species, all the founders of arbitrary governments and false religions have been extraordinary people; and nine tenths of the calamities that have befallen the human race had no other origin than the union of high intelligence with low desires.

- Thomas Macaulay

✧✧✧

A good soldier, whether he leads a platoon or an army, is expected to look backward as well as forward; but he must think only forward.

- General of the Army Douglas MacArthur

✧✧✧

In no other profession are the penalties for employing untrained personnel so appalling or so irrevocable as in the military.

- General Douglas MacArthur

✧✧✧

The history of failure in war can be summed up in two words: too late. Too late in comprehending the deadly purpose of a potential enemy; too late in realizing the mortal danger; too late in preparedness; too late in uniting all possible forces for resistance; too late in standing up with one's friends.

- General of the Army Douglas MacArthur

✧✧✧

It is my earnest hope—indeed the hope of all mankind—that from this solemn occasion a better world shall emerge out of the blood and carnage of the past, a world found upon faith and understanding, a world dedicated to the dignity of man and the fulfilment of his most cherished wish for freedom, tolerance and justice.

- General Douglas MacArthur
Supreme Allied Commander of South-West Pacific (1945)

✧✧✧

I Shall Return.

- Gen Douglas MacArthur

✧✧✧

There is no security in this life. There is only opportunity.

- Gen Douglas MacArthur

✧✧✧

In war, you win or lose, live or die—and the difference is an eyelash.

- General Douglas MacArthur

✧✧✧

In war there is no substitute for victory.

- General Douglas MacArthur

✧✧✧

Part of the American dream is to live long and die young. Only those Americans who are willing to die for their country are fit to live.

- General Douglas MacArthur

✧✧✧

The powers in charge keep us in a perpetual state of fear keep us in a continuous stampede of patriotic fervour with the cry of grave national emergency. Always there has been some terrible evil to gobble us up if we did not blindly rally behind it by furnishing the exorbitant sums demanded. Yet, in retrospect, these disasters seem never to have happened, seem never to have been quite real.

- General Douglas MacArthur

✧✧✧

We are not retreating—we are advancing in another direction.

- General Douglas MacArthur

✧✧✧

By profession I am a soldier and take pride in that fact. But I am prouder– infinitely prouder – to be a father. A soldier destroys in order to build; the father only builds, never destroys. The one has the potentiality of death; the other embodies creation and life. And while the hordes of death are mighty, the battalions of life are mightier still. It is my hope that my son, when I am gone, will remember me not from the battle field but in the home repeating with him our simple daily prayer, "Our Father Who Art in Heaven."

- General Douglas MacArthur

✧✧✧

No man is entitled to the blessings of freedom unless he be vigilant in its preservation.

- General Douglas MacArthur

✧✧✧

A general is just as good or just as bad as the troops under his command make him.

- General Douglas MacArthur

✧✧✧

The best luck of all is the luck you make for yourself.

- General Douglas MacArthur

✧✧✧

Every nation that would preserve its tranquillity, its riches, its independence and its self-respect must keep alive its martial ardour and be at all time prepared to defend itself.

- General Douglas MacArthur
(speech to the Rainbow Division, 14 July 1935)

✧✧✧

The trouble with opportunity is that it always comes disguised as hard work.

- General Douglas MacArthur

✧✧✧

In many situations that seemed desperate, the artillery has been a most vital factor.

- General Douglas MacArthur

✧✧✧

You are remembered for the rules you break.

- General Douglas MacArthur

✧✧✧

Age wrinkles the body. Quitting wrinkles the soul.

- General Douglas MacArthur

✧✧✧

I have known war as few men now living know it. It's very destructiveness on both friend and foe has rendered it useless as a means of settling international disputes.

- General Douglas MacArthur

✧✧✧

It is fatal to enter any war without the will to win it.

- General Douglas MacArthur

✧✧✧

The soldier, above all other people, prays for peace, for he must suffer and bear the deepest wounds and scars of war.

- General Douglas MacArthur

✧✧✧

One cannot wage war under present conditions without the support of public opinion, which is tremendously moulded by the press and other forms of propaganda.

- General Douglas MacArthur

✧✧✧

On the fields of friendly strife are sown the seed that on other days and other fields will bear the fruits of victory.

- General Douglas MacArthur

✧✧✧

Old soldiers never die; they just fade away. And like the old soldier in that ballad, I now close my military career and just fade away, an old soldier who tried to do his duty as God gave him the sight to see that duty.

- General Douglas MacArthur

✧✧✧

We must hold our minds alert and receptive to the application of unglimpsed methods and weapons. The next war will be won in the future, not in the past. We must go on, or we will go under.

- General Douglas MacArthur (as Chief of Staff, 1931)

✧✧✧

Sometimes it is suggested that Canada does not need armed forces in peacetime because if war comes we will be able to find the necessary experts. It is true that we can recruit doctors, engineers and the logisticians from civilian life. However, what is not understood is that we cannot find and we cannot hire from any civilian profession men who are skilled in the art of leading and training men for war. This is the special expertise possessed only by those of you who are trained in the art of military leadership. Men like you cannot be hired, they must be grown and educated in Canada, in peacetime.

- Major General Bruce F. Macdonald
Address at Combat Arms School, August 1976

✧✧✧

A prince should therefore have no other aim or thought, nor take up any other thing for his study but war and it organization and discipline, for that is the only art that is necessary to one who commands.

- Niccolo Machiavelli (1469–1527)
The Prince

✧✧✧

The first method of estimating the intelligence of a ruler is to look at the men he has around him.

- Niccolo Machiavelli
The Prince

✧✧✧

There is no avoiding war; it can only be postponed to the advantage of others.

- Niccolò Machiavelli (1469–1527)
Italian political philosopher, statesman.

✧✧✧

I hold it to be of great prudence for men to abstain from threats and insulting words towards any one, for neither the one nor the other in any way diminishes the strength of the enemy; but the one makes him more cautious, and the other increases his hatred of you, and makes him more persevering in his efforts to injure you.

- Niccolo Machiavelli

✧✧✧

You must know that there are two methods of fighting, one by law, the other by force; the first method is that of men, the second of beasts; but as the first method is often insufficient, one must have recourse to the second. It is, therefore, necessary for a prince to know well how to use both the beast and the man.

- Niccolo Machiavelli, The Prince

✧✧✧

War should be the only study of a prince. He should consider peace only as a breathing-time, which gives him leisure to contrive, and furnishes as ability to execute, military plans.

- Niccolo Machiavelli

✧✧✧

A prince never lacks legitimate reasons to break his promise.

- Niccolo Machiavelli

✧✧✧

Before all else, be armed.

- Niccolo Machiavelli

✧✧✧

It is double pleasure to deceive the deceiver.

- Niccolo Machiavelli

✧✧✧

Men are so simple and so much inclined to obey immediate needs that a deceiver will never lack victims for his deceptions.

- Niccolo Machiavelli

✧✧✧

Never was anything great achieved without danger.

- Niccolo Machiavelli

✧✧✧

No enterprise is more likely to succeed than one concealed from the enemy until it is ripe for execution.

- Niccolo Machiavelli

✧✧✧

The main foundations of every state, new states as well as ancient or composite ones, are good laws and good arms you cannot have good laws without good arms, and where there are good arms, good laws inevitably follow.

- Niccolo Machiavelli

✧✧✧

There is nothing more difficult to take in hand, more perilous to conduct, or more uncertain in its success, than to take the lead in the introduction of a new order of things.

- Niccolo Machiavelli

✧✧✧

If an injury has to be done to a man it should be so severe that his vengeance need not be feared.

- Niccolo Machiavelli

✧✧✧

The lion cannot protect himself from traps, and the fox cannot defend himself from wolves. One must therefore be a fox to recognize traps, and a lion to frighten wolves.

- Niccolo Machiavelli

✧✧✧

Men are driven by two principal impulses, either by love or by fear.

- Niccolo Machiavelli

✧✧✧

It is not titles that honour men, but men that honour titles.

- Niccolo Machiavelli

✧✧✧

From this arises the following question: whether it is better to be loved than feared, or the reverse. The answer is that one would like to be both the one and the other, but because they are difficult to combine, it is far better to be loved than feared if you cannot be both.

- Niccolo Machiavelli

✧✧✧

Occasionally words must serve to veil the facts. But let this happen in such a way that no one become aware of it; or, if it should be noticed, excuses must be at hand to be produced immediately.

- Niccolo Machiavelli

✧✧✧

Upon this, one has to remark that men ought either to be well treated or crushed, because they can avenge themselves of lighter injuries, of more serious ones they cannot; therefore the injury that is to be done to a man ought to be of such a kind that one does not stand in fear of revenge.

- Niccolo Machiavelli

✧✧✧

The Romans never allowed a trouble spot to remain simply to avoid going to war over it, because they knew that wars don't just go away, they are only postponed to someone else's advantage.

- Niccolo Machiavelli

✧✧✧

A prince ought to have no other aim or thought, nor select anything else for his study, than war and its rules and discipline; for this is the sole art that belongs to him who rules, and it is of such force that it not only upholds those who are born princes, but it often enables men to rise from a private station to that rank.

- Niccolo Machiavelli

✧✧✧

Whenever men are not obliged to fight from necessity, they fight from ambition; which is so powerful in human breasts, that it never leaves them no matter to what rank they rise. The reason is that nature has so created men that they are able to desire everything but are not able to attain everything: so that the desire being always greater than the acquisition, there results discontent with the possession and little satisfaction to themselves from it.

- Niccolo Machiavelli

✧✧✧

War is just when it is necessary; arms are permissible when there is no hope except in arms.

- Niccolo Machiavelli

✧✧✧

Good young officers who become good old generals are made by good sergeants, ... a combination of ill-founded self-confidence, bluff and outstanding support and guidance from a series of unforgettable sergeants allowed me to create an impression of competence.

- Major General Lewis MacKenzie

✧✧✧

A large number of our non-commissioned officers had done multiple tours in Cyprus, and a few had been there six or seven times. I had an excellent regimental sergeant-major, Bill Colbourne, whom I called into my office. "RSM, with all the soldiers we're going to leave behind, why don't we give some of the sergeants and warrant officers who've been there many times an opportunity to volunteer? They've paid their peacekeeping dues. They should only have to go back if they really want to."

"Right on, sir. I'll get the word out this afternoon," replied the RSM.

The following morning, when I arrived at my office, I found a line of sergeants and warrant officers extending from inside the front entrance of our headquarters building out onto the sidewalk.

The RSM was waiting for me in my office. "Sir, we have a problem," he explained. "Every one of those warrant officers and sergeants lined up in front of your office has the same story. They all want to go to Cyprus. But if their wives find out they volunteered, they will never hear the end of it. In some cases, I wouldn't be surprised if it breaks up the marriage. They all respectfully request that you order them to go to Cyprus!"

And so my dumb idea regarding volunteers was cancelled; we ordered most of the soldiers in the RSM'S line-up to Cyprus, and I filed away another lesson in human nature.

- Major-General Lewis MacKenzie, Peacekeeper, 1993

✧✧✧

Who rules East Europe commands the Heartland;
Who rules the Heartland commands the World Island;
Who rules the World Island commands the World.

- Halford Mackinder

This message was composed to convince the world statesmen at the Paris Peace conference, in 1919, of the crucial importance of Eastern Europe as the strategic route to the Heartland was interpreted as requiring a strip of buffer state to separate Germany and Russia.

✧✧✧

Knowledge is one. Its division into subjects is a concession to human weakness.

- Halford Mackinder

✧✧✧

A general must be shot or befriended—but never hurt.

- Salvador de Madariaga, Spanish politician and historian

✧✧✧

The future doesn't belong to the fainthearted; it belongs to the brave... We will never forget them, nor the last time we saw them, this morning, as they prepared for their journey and waved goodbye and 'slipped the surly bonds of Earth' to 'touch the face of God.

- John Gillespie Magee
1941 poem 'High Flight'

✧✧✧

Arjun chose Krishna, though Krishna had vowed to lay down his arms on the battlefield ... ... Krishna asked: "Why did you pick me, knowing I would not fight?" "I can handle the soldiers myself, O Krishna, if I have your presence to give me moral support. Some of your glory will surely rub off on me."

- The Mahabharata

✧✧✧

When the Gods deal defeat to a person, they first take his mind away, so that he sees things wrongly. .... Time does not raise a stick and clobber a man's head; the power of Time is just this upended view of things.

- The Mahabharata, Dhritarastra,
(The Book of the Assembly Hall)

✧✧✧

Revenge is not always better, but neither is forgiveness; learn to know them both, son, so that there is no problem.

- The Mahabharata, Prahlada
(quoted by Draupadi) (The Book of the Forest)

✧✧✧

I do not blame you, Maharaja, for hitting an innocent man. For, cruelty comes quick to the powerful.

- The Mahabharata, Yudhisthira to Virata
(The Book of Virata)

✧✧✧

Once war has been undertaken, no peace is made by pretending there is no war.

- Mahabharata Duryodhana (The Book of the Effort)

✧✧✧

Cleverness does not always lead to gain nor stupidity to poverty.

- The Mahabharata ,Vidura
(The Book of the Effort)

✧✧✧

You have the right to work, but for the work's sake only. You have no right to the fruits of work. Desire for the fruits of work must never be your motive in working. Never give way to laziness, either.

Perform every action with your heart fixed on the Supreme Lord. Renounce attachment to the fruits. Be even tempered in success and failure; for it is this evenness of temper which is meant by yoga.

Work done with anxiety about results is far inferior to work done without such anxiety, in the calm of self-surrender. Seek refuge in the knowledge of Brahman. They who work selfishly for results are miserable.

- *The Mahabharata (Bhagavadgita)*

✧✧✧

The soul can never be cut to pieces by any weapon, nor burned by any fire, nor moistened by water, nor withered by the wind."

- *The Mahabharata (Bhagavadgita)*

✧✧✧

There is neither this world nor the world beyond nor happiness for the one who doubts.

- *The Mahabharata (Bhagavadgita)*

✧✧✧

Delusion arises from anger. The mind is bewildered by delusion. Reasoning is destroyed when the mind is bewildered. One falls down when reasoning is destroyed.

- *The Mahabharata (Bhagavadgita)*

✧✧✧

The mind acts like an enemy for those who do not control it.

- *The Mahabharata (Bhagavadgita)*

✧✧✧

Man is made by his belief. As he believes, so he is.

- *The Mahabharata (Bhagavadgita)*

✧✧✧

In battle, in the forest, at the precipice in the mountains, on the dark great sea, in the midst of javelins and arrows, In sleep, in confusion, in the depths of shame, The good deeds a man has done before defend him.

- *The Mahabharata (Bhagavadgita)*

✧✧✧

A man's own self is his friend. A man's own self is his foe.

- *The Mahabharata (Bhagavadgita)*

✧✧✧

Death is as sure for that which is born, as birth is for that which is dead. Therefore grieve not for what is inevitable.

- *The Mahabharata (Bhagavadgita)*

✧✧✧

Valour, glory, firmness, skill, generosity, steadiness in battle and ability to rule—these constitute the duty of a soldier. They flow from his own nature.

- The Mahabharata (Bhagavadgita)

✧✧✧

This is the sum of all true righteousness: deal with others as thou wouldst thyself be dealt by.

Do nothing to thy neighbour which thou wouldst not have him do to thee hereafter.

- The Mahabharata (Bhagavadgita)

✧✧✧

Those far distant, storm-beaten ships, upon which the Grand Army never looked, stood between it and the dominion of the world.

- Alfred Thayer Mahan (1840-1914): The Influence of Sea Power upon the French Revolution and Empire, 1793–1812 (1892)

✧✧✧

The considerations and principles which enter into them belong to the unchangeable, or unchanging, order of things, remaining the same, in cause and effect, from age to age. They belong, as it were, to the Order of Nature, of whose stability so much is heard in our day; whereas tactics, using as its instruments the weapons made by man, shares in the change and progress of the race from generation to generation. From time to time the superstructure of tactics has to be altered or wholly torn down; but the old foundations of strategy so far remain, as though laid upon a rock.

- Alfred Thayer Mahan (1840–1914)
The Influence of Sea Power Upon History 1660–1783

✧✧✧

Finally, it must be remembered that, among all changes, the nature of man remains much the same; the personal equation, though uncertain in quantity and quality in the particular instance, is sure always to be found.

- Alfred Thayer Mahan (1840–1914)
The Influence of Sea Power Upon History 1660–1783

✧✧✧

The allies we gain by victory will turn against us upon the bare whisper of our defeat.

- Rear Admiral Alfred Thayer Mahan, Naval Historian 1911

✧✧✧

Free supplies and open retreat are two essentials to the safety of an army or a fleet.

- Rear Admiral Mahan

✧✧✧

Defeat cries loud for explanation; whereas success, like charity, covers a multitude of sins.

- Alfred Thayer Mahan

✧✧✧

Lesson ... is the danger of disgraceful failure to men who have neglected to keep themselves prepared, not only in knowledge of their profession, but in the sentiment of what war requires.

- Rear Admiral A.T. Mahan
The Influence of Seapower Upon History, 1660-1783

✧✧✧

Whether a democratic government will have the foresight, the keen sensitivity to the national position...to ensure its prosperity by adequate outpouring of money in times of peace, all of which are necessary for military preparation, is yet an open question.

- Rear Admiral A.T. Mahan
The Influence of Seapower Upon History, 1660-1783

✧✧✧

When the sea not only borders or surrounds but also separates a country into two or more parts, the control of it becomes not only desirable but vitally necessary.

- Rear Admiral A.T. Mahan
The Influence of Seapower Upon History, 1660–1783

✧✧✧

Only destruction of the enemy can be called victory.

- Admiral Stepan O. Makarov, 1904, Port Arthur

✧✧✧

To command is to serve, nothing more and nothing less.

- Andre Malraux (Man's Hope)

✧✧✧

A good head and a good heart are always a formidable combination.

- Nelson Mandela

✧✧✧

I was called a terrorist yesterday, but when I came out of jail, many people embraced me, including my enemies, and that is what I normally tell other people who say those who are struggling for liberation in their country are terrorists. I tell them that I was also a terrorist yesterday, but, today, I am admired by the very people who said I was one.

- Nelson Mandela, 16 May 2000

✧✧✧

I wonder whether those of our political masters who have been put in charge of the defence of the country can distinguish a mortar from a motor; a gun from a howitzer; a guerrilla from a gorilla, although a great many resemble the latter.

- Field Marshal Sam Manekshaw

✧✧✧

If anyone tells you he is never afraid, he is a liar or a Gurkha!

- Field Marshal Sam Manekshaw

✧✧✧

Gentleman, I have arrived! There will be no more withdrawals in 4 Corps, thank you.

- Field Marshal Sam Manekshaw
(on assumption of command of 4 Corps, 28 November 1962)

✧✧✧

A 'yes' man is a dangerous man.

- Field Marshal Sam Manekshaw

✧✧✧

You received three; when I was of your age, I received nine bullets and look today I am the Commander in Chief of the Indian Army.

- Field Marshal Sam Manekshaw
(During the 1971 Indo-Pakistan War when he met an injured soldier in Army Hospital with three bullet wounds.)

✧✧✧

You are going to be given command of troops in an operational area. Your task will be to administer the needs and to lead them in battle. What sort of men will you be commanding? You will be leading and dealing with veterans? You will be leading those who have fought wars, men who have won wars, men who are used to good things. Make sure you give them the best of leadership. Make it very clear to them, what you expect from them and also what you can give them.

- Field Marshal Sam Manekshaw
(Speech to gentlemen cadets at IMA, Dehradun on 30 March 1972)

✧✧✧

What is moral courage? It is the ability to distinguish right from wrong and having so distinguished it, be prepared to say so, irrespective of the views held by your superiors or subordinates and of consequence to yourself.

- Field Marshal Sam Manekshaw

✧✧✧

Discipline is the code of conduct of decent living.

- Field Marshal Sam Manekshaw

✧✧✧

To those of my commanders, who took an inordinately long time to come to a decision—I coined a Manekshawism "If you have to be a bloody fool, be one quickly".

- Field Marshal Sam Manekshaw

✧✧✧

"Don't you think I would be a worthy replacement for you, madam prime minister? You have a long nose. So have I. But I don't poke my nose into other people's affairs..."

- Field Marshal Sam Manekshaw

✧✧✧

Act like a man of thought—Think like a man of action.

- Thomas Mann

✧✧✧

For a strong adversary (corps) the opposition of twenty-four squadrons and twelve guns ought not to have appeared very serious, but in war the psychological factors are often decisive. An adversary who feels inferior is in reality so.

- Field Marshal Carl Gustav Baron Von Mannerheim, 1953

✧✧✧

Borders are scratched across the hearts of men
By strangers with a calm, judicial pen,
And when the borders bleed we watch with dread
The lines of ink across the map turn red.

- Marya Mannes, Subverse: Rhymes for Our Times, 1959

✧✧✧

Being a brilliant military commander is about more than just battlefield nous. It is about mobility and manoeuvre; half the battle is being in the right place at the right time.

- The Duke of Marlborough, 1703

✧✧✧

Two main charges were brought to the House of Commons against Marlborough: first, an assertion that over nine years he had illegally received more than £63,000 from the bread and transport contractors in the Netherlands; second, that he had taken 2.5% from the pay of the foreign troops in English pay, amounting to £280,000." were specious, as was demonstrated when his successor Ormonde took command of the army.

When the Duke of Ormonde, left London for The Hague to take command of British forces he went with 'the same allowances that had been voted criminal in the Duke of Marlborough.

You do well apprehend that good order and military discipline are the chief essentials in an army. But you must ever be aware that an army cannot preserve good order unless its soldiers have meat in their bellies, coats on their backs and shoes on their feet. All these are as necessary as arms and munitions. I pray you will never fail to look to these things as you may do to other matters.

- Marlborough, to his Quartermaster-General December 1703

✧✧✧

The warrior's gift is to willingly storm Hell that Heaven may remain unstained.

- R.V.A. Marcell

✧✧✧

Terrorism isn't James Bond or Tom Clancy. Even al-Qaeda is looking old school these days-now it's just some guy with a bomb. He walks the same roads as us. He thinks the same thoughts. But he's got a bomb.

- Michael Marshall, Blood of Angels

✧✧✧

Stonewall Jackson would rather lose one man to hard marching, than lose five men to hard battle. Perspiration saves blood!

- Colonel Marttinen
(to his tired and battle weary men in Infantry Regiment 61)

✧✧✧

Patriotism...is the egg from which wars are hatched.

- Guy de Maupassant

✧✧✧

Every government has as much of a duty to avoid war as a ship's captain has to avoid a shipwreck.

- Guy de Maupassant

✧✧✧

The safety of an armoured formation in the enemy's rear depends on its continued movement.

- Field Marshal Erich von Manstein

✧✧✧

Whenever things go well, news gets back quickly enough. If, on the other hand, an attack gets stuck, a blanket of silence descends on the front, either because communications are cut or those concerned prefer to hang on till they have something encouraging to report.

- Field Marshal Erich von Manstein

✧✧✧

There are only four types of officers. First, there are the lazy, stupid ones. Leave them alone, they do no harm...Second, there are the hard-working, intelligent ones. They make excellent staff officers, ensuring that every detail is properly considered. Third, there are the hard-working, stupid ones. These people are a menace and must be fired at once. They create irrelevant work for everybody. Finally, there are the intelligent, lazy ones. They are suited for the highest office.

- General Erich Von Manstein (1887–1973)
on the German Officer Corps

✧✧✧

Strength of character and inner fortitude, however, are decisive factors. The confidence of the man in the ranks rests upon a man's strength of character.

- Field Marshal Erich von Manstein

✧✧✧

War is the severest test of spiritual and bodily strength. In war, character outweighs intellect. Many stand forth in the field of battle who in peace would remain unnoticed.

- German Army Field Manual

✧✧✧

When a general complains of the morale of his troops, the time has come to look at his own.

- General George C. Marshall (1880–1959)
Chief of Staff of the US Army, Secretary of State

✧✧✧

When you are commanding, leading [soldiers] under conditions where physical exhaustion and privations must be ignored, where the lives of [soldiers] may be sacrificed, then, the efficiency of your leadership will depend only to a minor degree on your tactical ability. It will primarily be determined by your character, your reputation, not much for courage-which will be accepted as a matter of course-but by the previous reputation you have established for fairness, for that high-minded patriotic purpose, that quality of unswerving determination to carry through any military task assigned to you.

- General of the Army George C. Marshall, Speaking to officer candidates in September, 1941

✧✧✧

We are determined that before the sun sets on this terrible struggle our flag will be recognized throughout the world as a symbol of freedom on the one hand, of overwhelming power on the other.

- George C. Marshall, Chief of Staff

✧✧✧

There is no limit to the good you can do if you don't care who gets the credit.

- General George C. Marshall

✧✧✧

A political problem thought of in military terms eventually becomes a military problem.

- George C. Marshall

✧✧✧

We are determined that before the sun sets on this terrible struggle our flag will be recognized throughout the world as a symbol of freedom on the one hand, of overwhelming power on the other.

- General George C. Marshall

✧✧✧

The time has come when we must proceed with the business of carrying the war to the enemy, not permitting the greater portion of our armed forces and our valuable material to be immobilized within the continental United States.

- General George C. Marshall, 3 March 1942

✧✧✧

The one great element in continuing the success of an offensive is maintaining the momentum.

- General George C. Marshall

✧✧✧

If man does find the solution for world peace it will be the most revolutionary reversal of his record we have ever known.

- General George C. Marshall

✧✧✧

It is not enough to fight. It is the spirit which we bring to the fight that decides the issue. It is morale that wins the victory.

- General George C. Marshall

✧✧✧

Don't fight the problem, decide it.

- General George C. Marshall

✧✧✧

Military power wins battles, but spiritual power wins wars.

- General George C. Marshall

✧✧✧

Control presupposes that the leaders knows the location of all elements of his command at all times and can communicate with any element at any time.

- General George C. Marshall

(this sentiment, necessary for readying an unprepared amateur force for war, later helped foster the culture of micromanagement in the U.S. military)

✧✧✧

Also remember that in any man's dark hour, a pat on the back and an earnest handclasp may work a small miracle.

- Brigadier-General S.L.A Marshall,
The Armed Forces Officer 1950

✧✧✧

The starting point for the understanding of war is the understanding of human nature.

- S.L.A. Marshall, Men Against Fire

✧✧✧

Studies by Medical Corps psychiatrists of combat fatigue cases... found that fear of killing, rather than fear of being killed, was the most common cause of battle failure, and that fear of failure ran a strong second.

- S.L.A. Marshall, Men Against Fire

✧✧✧

In the Pacific campaigns, our forces were impressed time after time by the phenomenon of enemy troops (Japanese) who had quit their arms and who appeared incapable of any offensive or self-protecting gesture. Yet these troops stood their ground like plants rooted in the earth and insisted on being killed to the last man. Their living bodies were the defensive base around which the action of their more willing comrades proceeded.

- S.L.A. Marshall, Men Against Fire

✧✧✧

Battles are won through the ability of men to express concrete ideas in clear and unmistakable language.

- S.L.A. Marshall, Men Against Fire

✧✧✧

In training, there being no real bullet danger even on the courses which employ live ammunition, every advance under a supposed enemy fire is unrealistic.

- S.L.A. Marshall, Men Against Fire

✧✧✧

A man has integrity if his interest in the good of the service is at all times greater than his own personal pride, and when he holds himself to the same line of duty when unobserved as he would follow if his superiors were present.

- S.L.A. Marshall, Men Against Fire

✧✧✧

The art of leading, in operations large or small, is the art of dealing with humanity, of working diligently on behalf of men, of being sympathetic with them, but equally, of insisting that they make a square facing toward their own problems.

- S.L.A. Marshall, Men Against Fire

✧✧✧

Artillerymen have a love for their guns which is perhaps stronger than the feeling of any soldier for his weapon or any part of his equipment.

- Brigadier General S. L. A. Marshall

✧✧✧

The salient characteristics of most of our great victories (and a few of our defeats) was that they pivoted on the fire action of a few men.

- S.L.A. Marshall, Men Against Fire

✧✧✧

However much we may honour the "Unknown Soldier" as the symbol of sacrifice in war, let us not mistake the fact that it is the "Known Soldier" who wins battles. Sentiment aside, it is the man whose identity is well known to his fellows who has the main chance as a battle effective.

- S.L.A. Marshall, Men Against Fire: the Problem of Battle Command in Future War, 1947

✧✧✧

Solidarity with the group was one of the most important motivations for soldiers in battle: I hold it to be one of the simplest truths of war that the thing which enables an infantry soldier to keep going with his weapons is the near presence or the presumed presence of a comrade.

- S.L.A. Marshall, Men Against Fire

✧✧✧

Fear is general among men [but] men are commonly loath that their fear will be expressed in specific acts which their comrades will recognize as cowardice.

- S.L.A. Marshall, Men Against Fire

✧✧✧

In the whole of the initial assault landings on Omaha beachhead, there were only about five infantry companies which were tactically effective during the greater part of 6 June 1944. In these particular companies, an average of about one-fifth of the men fired their weapons during the day-long advance from the water's edge to the first tier of villages inland—a total of perhaps not more than 450 men firing consistently with infantry weapons in the decisive companies. These facts were determined by a systematic check of the survivors. It was not a story of great volume, even for the men who fired. Only one company was able to unite a base of fire for any period. The company which made the deepest penetration losing a high percentage of its men in so doing, saw only six 'live Germans' during its advance, and these turned out to be Russians. The day was conspicuous for its lack of live targets.

- S.L.A. Marshall, Men Against Fire

✧✧✧

When an advancing infantry line suddenly encounters enemy fire and the men go to ground under circumstances where they cannot see one another, the moral disintegration of that line is, for the moment, complete. All organization unity vanishes temporarily. What has been a force becomes a scattering of individuals. This is inevitably the case. Men going forward in line are insight of one another. They, therefore, have a sense of unity. But when they come unexpectedly in check and go to ground, they no longer have knowledge of the position of the men on left and right because they have not taken true cognizance of these things as they moved along. Indeed, it is not possible for them to do so if they are to be alert to the danger which lies ahead. While erect, they feel the presence of others; when they go down this feeling is lost.

- S.L.A. Marshall, Men Against Fire

✧✧✧

Leaders must talk if they are to lead. Action is not enough. A silent example will never rally men.

- S.L.A. Marshall, Men Against Fire

✧✧✧

We have encouraged the man to think creatively as a person without stimulating him to act and speak at all times as a member of a team. His first duty is to join his force to others! Squad unity comes to full cooperation between each man and his neighbour. There is not battle strength with the company or regiment except as it derives from this basic element within the smallest component.

- S.L.A. Marshall, Men Against Fire

✧✧✧

"...the search for information and the giving of it are the true beginnings of what is called initiative."

- S.L.A. Marshall, Men Against Fire

✧✧✧

It is the soldier acting on his own to advise others of his tactical situation or conveying any other information which may be of general benefit in furthering the tactical situation of the company or in enlisting the aid of others in carrying out any action which will benefit the tactical situation of the company.

- S.L.A. Marshall, Men Against Fire

✧✧✧

In the majority of men the retention of self-discipline under conditions of the battlefield depends upon the maintaining of an appearance of discipline within the unit. Should the latter begin to dissolve, only a small minority of the most hardy individuals will retain self- control.

- S.L.A. Marshall, Men Against Fire

✧✧✧

If you only have a hammer, you tend to see every problem as a nail.

- Abraham Maslow

✧✧✧

One's only rival is one's own potentialities. One's only failure is failing to live up to one's own possibilities. In this sense, every man can be a king, and must therefore be treated like a king.

- Abraham Maslow

✧✧✧

Sighted sub, sank same.

- Enlisted pilot Donald F. Mason
radio dispatch, Jan. 28, 1942. He believed that
he had sunk a German U-boat off Argentina, Newfoundland.

✧✧✧

What they saw in India confirmed their views; there were hereditary rulers, soldiers, moneylenders, washermen, goldsmiths and scavengers. It was common sense to recognize this and not waste time or effort trying to turn moneylenders into soldiers. Besides, to recognize the difference was the surest way of enlisting the loyalty of the 'martial races' as they came to be called. The division of people into 'martial' and 'non-martial' classes was not an invention of the British but it was the recognition of something already existing in the Indian social system. But it was extremely convenient to a conqueror.

- Philip Mason (A Matter of Honour)

✧✧✧

Do you know what a soldier is, young man? He's the chap who makes it possible for civilized folk to despise war.

- Allan Massie

✧✧✧

Military justice is to justice what military music is to music.

- Groucho Marx

✧✧✧

Philosophers have only interpreted the world. The point, however, is to change it.

- Karl Marx

✧✧✧

Men make their own history, but they do not make it just as they please; they do not make it under circumstances chosen by themselves, but under circumstances directly encountered, given, and transmitted from the past.

- Karl Marx

✧✧✧

You can't learn too soon that the most useful thing about a principle is that it can always be sacrificed to expediency.

- W Somerset Maugham

✧✧✧

Nuts.

- Anthony McAuliffe

Acting commander of the 101st Airborne Division during the Battle of the Bulge in World War 11. He was in charge of the defence of Bastogne on December 22nd 1944 when the garrison was called on by advancing German forces to surrender.

✧✧✧

His initial response was 'Aw, Nuts!' When he came to compile a written reply he could think of nothing more appropriate:

'To the German Commander: NUTS!
The American Commander.

Bastogne was successfully held by the Americans and Anthony C. McAuliffe became immortalized for a single word.

Today the real test of power is not the capacity to make war but the capacity to prevent it.

- Anne O'Hare McCormick

✧✧✧

As an American asked to serve, I was prepared to fight, to be wounded, to be captured, and even prepared to die, but I was not prepared to be abandoned.

- Former POW Eugene Red McDaniel

✧✧✧

Modern war has become a struggle for men's minds as well as for their bodies.

- Brigadier General Robert McClure, Korean War

✧✧✧

Everybody in Afghanistan ought to know we're coming in and hells coming with us.

- Former National Security Advisor Robert McFarlane
(CBS Face the Nations, Oct. 28 2001.)

✧✧✧

IN FLANDERS FIELDS

In Flanders fields the poppies blow
Between the crosses, row on row,
That mark our place; and in the sky
The larks, still bravely singing, fly
Scarce heard amid the guns below.

We are the Dead. Short days ago
We lived, felt dawn, saw sunset glow,
Loved and were loved, and now we lie,
In Flanders fields.

Take up our quarrel with the foe:
To you from failing hands we throw
The torch; be yours to hold it high.
If ye break faith with us who die
We shall not sleep, though poppies grow
In Flanders fields.

- John McCrae

✧✧✧

It doesn't require any particular bravery to stand on the floor of the Senate and urge our boys in Vietnam to fight harder, and if this war mushrooms into a major conflict and a hundred thousand young Americans are killed, it won't be US Senators who die. It will be American soldiers who are too young to qualify for the senate

- George McGovern

✧✧✧

I'm fed up to the ears with old men dreaming up wars for young men to die in.

- George McGovern

✧✧✧

The bomber was not a direct product of circumstances; it was the result of a gradual realization of the cardinal value of aircraft.

- Air Marshal E.J. Kingston-McCloughry

✧✧✧

History shows that army campaigns in undeveloped countries have often involved waging war against natural obstacles, rather than against a foe.

- Air Marshal E.J. Kingston-McCloughry

✧✧✧

Television brought the brutality of war into the comfort of the living room. Vietnam was lost in the living rooms of America—not on the battlefields of Vietnam.

- Marshall McLuhan

✧✧✧

A typewriter is a means of transcribing thought, not expressing it.

- Marshall McLuhan

✧✧✧

As technology advances, it reverses the characteristics of every situation again and again. The age of automation is going to be the age of 'do it yourself.'

- Marshall McLuhan

✧✧✧

We become what we behold. We shape our tools and then our tools shape us.

- Marshall McLuhan

✧✧✧

We drive into the future using only our rear-view mirror.

- Marshall McLuhan

✧✧✧

Most of our assumptions have outlived their uselessness.

- Marshall McLuhan

✧✧✧

The medium is the message.

- Marshall McLuhan

✧✧✧

The book is an extension of the eye... clothing, an extension of the skin... electric circuitry, an extension of the central nervous system.

- Marshall McLuhan

✧✧✧

Today there is no longer any such thing as military strategy; there is only crisis management.

- Robert McNamara,
US Secretary of Defence

✧✧✧

1. Empathize with your enemy.
2. Rationality will not save us.
3. There's something beyond oneself.
4. Maximize efficiency.
5. Proportionality should be a guideline in war.
6. Get the data.
7. Belief and seeing are both often wrong.
8. Be prepared to re-examine your reasoning.
9. In order to do good, you may have to engage in evil.
10. Never say never.
11. You can't change human nature.

- Robert McNamara
The Fog of War: 11 Lessons

✧✧✧

Any military commander who is honest with himself, or with those he's speaking to, will admit that he has made mistakes in the application of military power. He's killed people unnecessarily—his own troops or other troops—through mistakes, through errors of judgment. A hundred, or thousands, or tens of thousands, maybe even a hundred thousand. But, he hasn't destroyed nations. And the conventional wisdom is don't make the same mistake twice, learn from your mistakes. And we all do. Maybe we make the same mistake three times, but hopefully not four or five. There will be no learning period with nuclear weapons. You make one mistake and you're going to destroy nations.

- Robert McNamara

✧✧✧

Brains are like hearts—they go where they are appreciated.

- Robert McNamara

✧✧✧

A computer does not substitute for judgment any more than a pencil substitutes for literacy. But writing without a pencil is no particular advantage.

- Robert McNamara

✧✧✧

Never answer the question that is asked of you, answer the question that you wish had been asked of you.

- Robert McNamara

✧✧✧

We see what we want to believe.

- Robert McNamara

✧✧✧

Our warfighting concept has to take account of the fact that almost nothing ever works right. As with the game of golf, our only real hope is to make smaller mistakes.

- Gen Merrill McPeak

✧✧✧

It is a disgrace that modern air forces are still shackled to a planning and execution cycle that lasts three days. We have out jets to a hot air balloon.

- Gen Merrill McPeak, USAF

✧✧✧

Never doubt that a small group of thoughtful, committed citizens can change the world. Indeed, it is the only thing that ever has.

- Margaret Mead

✧✧✧

Men are at war with each other because each man is at war with himself.

- Francis Meehan

✧✧✧

We have always said that in our war with the Arabs we had a secret weapon—no alternative.

- Prime Minister of Israel, Golda Meir, 3 October 1969

✧✧✧

Lo nislach lachem shegramtem lanu lahrog et bneihem.

We will not forgive you for making us kill your sons.

- Prime Minister Golda Meir addressing the Arab nations

✧✧✧

Every civilization finds it necessary to negotiate compromises with its values.

- Golda Meir (Prime Minister of Israel, 1969-1974)

✧✧✧

A leader who doesn't hesitate before he sends his nation into battle is not fit to be a leader.

- Golda Meir

✧✧✧

Authority poisons everybody who takes authority on himself.

- Golda Meir

✧✧✧

I can honestly say that I was never affected by the question of the success of an undertaking. If I felt it was the right thing to do, I was for it regardless of the possible outcome.

- Golda Meir

✧✧✧

Let me tell you something that we Israelis have against Moses. He took us 40 years through the desert in order to bring us to the one spot in the Middle East that has no oil!

- Golda Meir

✧✧✧

One cannot and must not try to erase the past merely because it does not fit the present.

- Golda Meir

✧✧✧

We will have peace with the Arabs when they love their children more than they hate us.

- Golda Meir

✧✧✧

The great prizes can be won only by speed, daring and manoeuvre.

- Field Marshal von Mellenthin

✧✧✧

Just as the diamond requires three properties for its formation—carbon, heat, and pressure—successful leaders require the interaction of three properties—character, knowledge, and application. Like carbon to the diamond, character is the basic quality of the leader. But as carbon alone does not create a diamond, neither can character alone create a leader. The diamond needs heat. Man needs knowledge, study, and preparation. The third property, pressure—acting in conjunction with carbon and heat—forms the diamond. Similarly, one's character, attended by knowledge, blooms through application to produce a leader.

- General Edward C. Meyer, Former US Army Chief of Staff

✧✧✧

The whole aim of practical politics is to keep the populace alarmed (and hence clamorous to be led to safety) by menacing it with an endless series of hobgoblins, all of them imaginary.

- H. L. Mencken

✧✧✧

The basic fact about human existence is not that it is a tragedy, but that it is a bore. It is not so much a war as an endless standing in line.

- Henry Louis Mencken

✧✧✧

War may make a fool of man, but it by no means degrades him; on the contrary, it tends to exalt him, and its net effects are much like those of motherhood on women.

- Henry L. Mencken

✧✧✧

A decision in the air must be sought and obtained before a decision on the ground can be reached.

- Billy Mitchell

✧✧✧

Nothing can stop the attack of aircraft except other aircraft.

- Billy Mitchell

✧✧✧

The advent of air power, which can go straight to the vital centres and either neutralize or destroy them, has put a completely new complexion on the old system of making war. It is now realized that the hostile main army in the field is a false objective, and the real objectives are the vital centres.

- Brigadier General William 'Billy' Mitchell, 'Skyways: A Book on Modern Aeronautics,' 1930.

✧✧✧

In the development of air power, one has to look ahead and not backward and figure out what is going to happen, not too much what has happened.

- Brigadier General William 'Billy' Mitchell, USAS.

✧✧✧

The most important branch of aviation is pursuit, which fights for and gains control of the air.

- Brigadier General William 'Billy' Mitchell, USAF.

✧✧✧

Changes in military systems come about only through the pressure of public opinion or disaster in war.

- Billy Mitchell (Winged Defence)

✧✧✧

The world stands on the threshold of the "aeronautical era." During this epoch the destinies of all people will be controlled through the air.

- Billy Mitchell (Winged Defence)

✧✧✧

The future of our nation is forever bound up in the development of Air Power.

- Colonel William 'Billy' Mitchell

✧✧✧

In the development of air power one has to look ahead and not backwards and figure out what is going to happen, not too much what has happened.

- Billy Mitchell (Winged Defence)

✧✧✧

Fighting is like champagne. It goes to the heads of cowards as quickly as of heroes. Any fool can be brave on a battlefield when it's be brave or else be killed.

- Margaret Mitchell

✧✧✧

War is an ugly thing, but not the ugliest of things. The decayed and degraded state of moral and patriotic feeling, which thinks that nothing is worth war, is much worse. The person who has nothing for which he is willing to fight, nothing which is more important than his own personal safety, is a miserable creature and has no chance of being free unless made and kept so by the exertions of better men than himself.

- John Stuart Mill (1868)

✧✧✧

Who overcomes by force, hath overcome but half his foe.

- Milton

✧✧✧

One cannot reflect concern downward.

- Maj Gen John O. Moench, USAF

✧✧✧

I am convinced that a bombing attack launched from such carriers [the *Lexington* and *Saratoga*] from an unknown point, at an unknown instant, with an unknown objective, cannot be warded off.

- Adm William A. Moffett, at the christening of the Lexington, 3 October 1925

✧✧✧

Never forget that a corpse never cares how it got to be so cold. Commanders should always keep in mind that they wage war through a wall of human blood, sweat, and tears whose pain they can never truly feel and whose loss they can never truly know. For they are become death: they are the destroyer of worlds.

- Ellen Mogensen

❖❖❖

First weigh the considerations, then take the risks.

- Helmuth von Moltke (1800–1891), German Field Marshal

❖❖❖

The tactical result of an engagement forms the base for new strategic decisions because victory or defeat in a battle changes the situation to such a degree that no human acumen is able to see beyond the first battle.

- Helmuth von Moltke

❖❖❖

No plan of operations extends with any certainty beyond the first contact with the main hostile force.

- Helmuth von Moltke

❖❖❖

Strategy is a system of expedients. It is more than science, it is the translation of science into practical life, the development of an original leading thought in accordance with the ever-changing circumstances.

- Helmuth von Moltke

❖❖❖

In the long run luck is given only to the efficient.

- Helmuth von Moltke

❖❖❖

War forms part of the order of things instituted by God.

- Helmuth von Moltke

❖❖❖

If one wishes to attack, then one must do so with resoluteness. Half measures are out of place. Only strength and confidence carry the units with them and produce success.

- Field Marshal Count Helmut von Moltke

❖❖❖

A favourable situation will never be exploited if commanders wait for orders. The highest commander and the youngest soldier must always be conscious of the fact that omission and inactivity are worse than resorting to the wrong expedient.

- Helmuth von Moltke

❖❖❖

The Army is the most outstanding institution in every country, for it alone makes possible the existence of all civic institutions.

- Field Marshal Helmut Graf von Moltke (1800-1891)

❖❖❖

The war of 1866 [between Prussia and Austria] was entered on not because the existence of Prussia was threatened, nor was it caused by public opinion and the voice of the people; it was a struggle, long foreseen and calmly prepared for, recognized as a necessity by the Cabinet, not for territorial aggrandizement, for an extension of our domain, or for material advantage, but for an ideal end—the establishment of power. Not a foot of land was exacted from Austria, but she had to renounce all part in the hegemony of Germany.... Austria had exhausted her strength in conquests south of the Alps, and left the western German provinces unprotected, instead of following the road pointed out by the Danube. Its centre of gravity lay out of Germany; Prussia's lay within it. Prussia felt itself called upon and strong enough to assume the leadership of the German races.

- Field Marshal Helmuth von Moltke

✧✧✧

Your Majesty! It cannot be done. The deployment of millions cannot be improvised. If Your Majesty insists on leading the whole army to the East it will not be an army ready for battle but a disorganized mob of armed men with no arrangements for supply. Those arrangements took a whole year of intricate labour to complete. .... Military plans dictate policy—and once settled, it cannot be altered.

- General Helmuth von Moltke, to the Kaiser
When the latter tried to persuade him to face the Russian front, instead of towards France at the start of the First World War

✧✧✧

We must be very careful what we do with the British infantry. Their fighting spirit is based largely on morale and regimental esprit de corps. On no account must anyone tamper with this.

- Lord Montgomery of Alamein

✧✧✧

Every soldier must know, before he goes into battle, how the little battle he is to fight fits into the larger picture, and how the success of his fighting will influence the battle as a whole.

- Field Marshal Bernard Law Montgomery

✧✧✧

Diplomats do not seem to understand that the game being played is poker, they play it as if it were chess.

- Field Marshal Viscount Montgomery

✧✧✧

I've got to go to meet God—and explain all those men I killed at Alamein.

- Field Marshall Viscount Montgomery, near the end of his life in 1976

✧✧✧

The morale of the soldier is the greatest single factor in war.

- Field Marshal Sir Bernard Law Montgomery

✧✧✧

The harder the fighting and the longer the war, the more the infantry, and in fact all the arms, lean on the gunners. The proper use of artillery is a great battle winning factor.

- Field Marshal B.L. Montgomery

✧✧✧

If we lose the war in the air we lose the war and lose it quickly.

- Field Marshall Bernard Montgomery

✧✧✧

Air power is indivisible. If you split it up into compartments, you merely pull it to pieces and destroy its greatest asset—its flexibility.

- Field Marshall Bernard Montgomery

✧✧✧

By the very nature of things, skill in the profession of arms has to be learned mostly in theory by studying the science of war—since the opportunity of practice in the art does not come often to the general. For this reason the great captains have always been serious students of military history...T.E. Lawrence rightly said that we of the twentieth century have two thousand years of experience behind us, and, if we still must fight, we have no excuse for not fighting well. My reading over the years has convinced me that nobody in this twentieth century can become a great commander, a supreme practitioner of the art of war, unless he has first studied and pondered its science.

- Viscount Montgomery of Alamein A History of Warfare

✧✧✧

I went to the Staff College at Camberley in January 1920 with no claim to cleverness. I thought I had a certain amount of common sense, but it was untrained; it seemed to me that it was trained common sense which mattered. I must admit that I was critical and intolerant; I had yet to learn that uninformed criticism is valueless.

- Field-Marshal Montgomery

✧✧✧

A "war against terrorism" is an impracticable conception if it means fighting terrorism with terrorism.

- John Mortimer, Where There's a Will...

✧✧✧

The strongest, most generous, and proudest of all virtues is true courage.

- Michel de Montagne, 1533–1592

✧✧✧

War hath no fury like a non-combatant.

- Charles Edward Montague

✧✧✧

The number of medals on an officer's breast varies in inverse proportion to the square of the distance of his duties from the front line.

- Charles Edward Montague

✧✧✧

It is a paradox to hope for victory without fighting. The goal of the man who makes war is to fight in the open field to win a victory.

- Field Marshal Prince Raimond Montecuccoli, 1703

✧✧✧

I will take care of the Spaniards in due time.

- Montezuma II (Moctezuma II) c1470–1520,
last Aztec emperor of Mexico 1502–20

✧✧✧

American soldiers in battle don't fight for what some presidents say on T.V., they don't fight for mom, apple pie, the American flag. They fight for one another.

- Col Hal Moore (7th Calvary, Vietnam)

✧✧✧

Leadership is the practical application of character.

- Colonel R. Meinertzhagen,
CBE, DSO, Army Diary, 1899-1926

✧✧✧

I believe that it is better in an emergency to do something not quite perfect than to sink into deep reflection or consult others and let the moment for action pass for ever. During my army career was told by a distinguished soldier that I "rushed my fences". Maybe I do, but I often surmount them.

- Colonel R. Meinertzhagen,
CBE, DSO, Army Diary, 1899–1926

✧✧✧

The whole aim of practical politics is to keep the populace alarmed (and hence clamorous to be led to safety) by menacing it with an endless series of hobgoblins, all of them imaginary.

- H. L. Mencken

✧✧✧

The basic fact about human existence is not that it is a tragedy, but that it is a bore. It is not so much a war as an endless standing in line.

- Henry Louis Mencken

✧✧✧

War may make a fool of man, but it by no means degrades him; on the contrary, it tends to exalt him, and its net effects are much like those of motherhood on women.

- Henry L. Mencken

✧✧✧

War is an ugly thing, but not the ugliest of things. The decayed and degraded state of moral and patriotic feeling which thinks that nothing is worth war is much worse. The person who has nothing for which he is willing to fight, nothing which is more important than his own personal safety, is a miserable creature, and has no chance of being free unless made or kept so by the exertions of better men than himself.

- John Stuart Mill

✧✧✧

Who overcomes by force, hath overcome but half his foe.

- Milton

✧✧✧

... then there was war in heaven. But it was not angels. It was that small golden zeppelin, like a long oval world, high up. It seemed as if the cosmic order were gone, as if there had come a new order, a new heavens above us: and as if the world in anger were trying to revoke it. ... So it seems ours cosmos is burst, burst at last, the stars and Moon blown away, the envelope of the sky burst out, and a new cosmos appeared, with a long-ovate, gleaming central luminary, calm and drifting in a glow of light, like a new Moon, with its light bursting in flashes on the earth, to burst away the earth also. So it is the end – our world is gone, and we are like dust in the air.

- Milton, 'Paradise Lost.'

✧✧✧

Never forget that a corpse never cares how it got to be so cold. Commanders should always keep in mind that they wage war through a wall of human blood, sweat, and tears whose pain they can never truly feel and whose loss they can never truly know. For they are become death: they are the destroyer of worlds.

- Ellen Mogensen

✧✧✧

The number of medals on an officer's breast varies in inverse proportion to the square of the distance of his duties from the front line.

- Charles Edward Montague

✧✧✧

War hath no fury like a non-combatant.

- Charles Edward Montague

✧✧✧

The strongest, most generous, and proudest of all virtues is true courage.

- Michel Eyquem de Montaigne

✧✧✧

If you don't know how to die, don't worry; Nature will tell you what to do on the spot, fully and adequately. She will do this job perfectly for you; don't bother your head about it.

- Michel de Montaigne

✧✧✧

If falsehood, like truth, had but one face, we would be more on equal terms. For we would consider the contrary of what the liar said to be certain. But the opposite of truth has a hundred thousand faces and an infinite field.

- Michel de Montaigne

✧✧✧

The strongest, most generous, and proudest of all virtues is true courage.

- Michel Eyquem de Montaigne, 1533–1592

✧✧✧

The air fleet of an enemy will never get within striking distance of our coast as long as our aircraft carriers are able to carry the preponderance of air power to sea.

- Real Admiral W. A. Moffet, Chief of the US Bureau of Aeronautics, October 1922.

✧✧✧

Without the grand design, informed by historic experience and seeking what is politically possible, foreign policy is blind; it moves without knowing where it is going.

- Hans J Morgenthau

✧✧✧

Aspiration for power is the distinguishing element of all politics.

- Hans J Morgenthau

✧✧✧

Political realism believes that politics, like society in general, is governed by objective laws that have their roots in human nature.

- Hans J Morgenthau

✧✧✧

The main signpost that helps political realism to find its way through the landscape of international politics is the concept of interest defined in terms of power.

- Hans J Morgenthau

✧✧✧

Realism assumes that its key concept of interest defined as power is an objective category which is universally valid, but it does not endow that concept with a meaning that is fixed once and for all.

- Hans J Morgenthau

✧✧✧

Traditionally "balance" of power denoted a policy by which one nation endeavoured...to counteract the power of another nation by increasing its strength [meaning chiefly, its military strength] to a point where it is at least equal, if not superior, to the other nation's strength.

- Hans J Morgenthau

✧✧✧

It is not only a political necessity but also a moral duty for a nation to follow in its dealings with other nations but one guiding star, one standard of thought, one rule of action: THE NATIONAL INTEREST.

- Hans J Morgenthau

✧✧✧

Political realism is aware of the moral significance of political action. It is also aware of the ineluctable tension between the moral command and the requirements of successful political action.

- Hans J Morgenthau

✧✧✧

The statesman must think in terms of the national interest, conceived as power among other powers. The popular mind, unaware of the fine distinctions of the statesman's thinking, reasons more often than not in the simple moralistic and legalistic terms of absolute good and absolute evil.

- Hans J Morgenthau

✧✧✧

The military services are particularly critical and repressive of the innovative types. The Italian Giulio Douhet, the apostle of total warfare, did a tour in prison for his heretical ideas. General 'Billy' Mitchell was tried, found guilty, and forced to resign from the Army—condemned on the charge of "making statements to the prejudice of good discipline". But fellow aviators know that the real reason for his trial and conviction was that his ideas differed from those accepted as doctrine by the War Department.

- Colonel James Mrazek (The Art of Winning Wars)

✧✧✧

If you wish to feign confusion in order to lure the enemy on, you must first have perfect discipline; if you wish to display timidity in order to entrap the enemy, you must have extreme courage; if you wish to parade your weakness in order to make the enemy over-confident, you must have exceeding strength.

- Tu Mu

✧✧✧

If I wish to wrest an advantage from the enemy, I must not fix my mind on that alone, but allow for the possibility of the enemy also doing some harm to me... If I wish to extricate myself from a dangerous position, I must consider not only the enemy's ability to injure me, but also my own ability to gain an advantage over the enemy.

- Tu Mu

✧✧✧

If the enemy is the invading party, we can cut his line of communications and occupy the roads by which he will have to return; if we are the invaders, we may direct our attack against the sovereign himself.

- Tu Mu

✧✧✧

When one tugs at a single thing in Nature, he finds it attached to the rest of the world.

- John Muir

✧✧✧

We're at a real time of transition here in terms of future aviation. What's going to be manned? What's going to be unmanned? There are those who see [the JSF] as the last manned fighter/bomber. And I'm one that's inclined to believe it—whether it's right or not.

- Admiral Michael Mullen, Joint Chiefs Chairman, congressional testimony regards the Fiscal 2010 defence budget and the future of manned military aviation, reported in Aviation Week & Space Technology, 18 May 2009.

✧✧✧

Courage is a morale quality: it is not a chance gift like aptitude for games. It is a cold choice between two alternatives, the fixed resolve not to quit; an act of renunciation which must be made not once but many times by the power of the will.

- Lord Moran, Anatomy of Courage

✧✧✧

Courage is will-power, whereof no man has an unlimited stock; and when in war it is used up, he is finished. A man's courage is his capital and he is always spending.

- Lord Moran

✧✧✧

He was taught the three arts of war, so much more necessary than musketry, field engineering or tactics. Or were they, perhaps, part of tactics? Wangling, Scrounging, and Winning. ... Wangling was the art of obtaining one's just due by unfair means. For instance, every officer and man of the B.E.F. had his allotted daily rations, his camp or billet, his turn for leave. In practice, to get these necessities, it was well to know the man who provided them and do him some small service-a bottle of whiskey, the loan of transport (if you had any) or of a fatigue party. Wangling extended to the lowest ranks. Men wangled from the N.C.O.s the better sorts of jam and extra turns off duty. ... Scrounging could be defined as obtaining that which one had not a shadow of a claim by unfair means. It was more insidious that as the Wangle, but just as necessary-men scrounged the best dug-outs off one another, or off neighbouring sections. N.C.O.s scrounged rum by keeping a thumb in the dipper while doling it out, Officers scrounged the best horse lines from other units. Colonials scrounged telephone wire to snare rabbits ... the Art of Winning. It may be defined as Stealing. More fully, it was the Art of obtaining that which one had no right to, for the sake of obtaining it, for the joy of possession. ... Some say it was simply the primeval joy of loot, ...

- R.H. Morrison, in Guy Chapman), Vain Glory; A miscellany of the Great War 1914-1918

✧✧✧

The preponderance of the Republican Guard divisions outside of Baghdad are now dead. I find it interesting when folks say we're softening them up. We're not softening them up, we're killing them.

- Lt. Gen. Michael Moseley, US Air Force
5 April 2003

✧✧✧

Friendly fire—isn't.
Recoilless rifles—aren't.
Suppressive fires—won't.
If it's stupid but it works, it isn't stupid.
Try to look unimportant; the enemy may be low on ammunition and not want to waste a bullet on you.
If at first you don't succeed, call in an air strike.
If you are forward of your position, your artillery will fall short.
Never share a foxhole with anyone braver than yourself.
Never forget that your weapon was made by the lowest bidder.
If your attack is going really well, it's an ambush.
The enemy diversion you're ignoring is their main attack.
The enemy invariably attacks on two occasions: when they're ready; and when you're not.
Five second fuses always burn three seconds.
There is no such thing as an atheist in a foxhole.
A retreating enemy is probably just falling back and regrouping.
The important things are always simple; the simple are always hard.
The easy way is always mined.
Don't look conspicuous; it draws fire.
If you are short of everything but the enemy, you are in the combat zone.
When you have secured the area, make sure the enemy knows it too.
Incoming fire has the right of way.
No combat ready unit has ever passed inspection.
If the enemy is within range, so are you.
The only thing more accurate than incoming enemy fire is incoming friendly fire.
Things that must work together, can't be carried to the field that way.
Radios will fail as soon as you need fire support.
Radar tends to fail at night and in bad weather, and especially during both.)
Anything you do can get you killed, including nothing.
Make it too tough for the enemy to get in, and you won't be able to get out.
Tracers work both ways.
When both sides are convinced they're about to lose, they're both right.
Professional soldiers are predictable; the world is full of dangerous amateurs.
Mines are equal opportunity weapons.
The one item you need is always in short supply.

Interchangeable parts aren't.

It's not the one with your name on it; it's the one addressed "to whom it may concern" you've got to think about.

When in doubt, empty your magazine.

The side with the simplest uniforms wins.

Combat will occur on the ground between two adjoining maps.

Never stand when you can sit, never sit when you can lie down, never stay awake when you can sleep.

The more a weapon costs, the farther you will have to send it away to be repaired.

The complexity of a weapon is inversely proportional to the IQ of the weapon's operator.

No matter which way you have to march, its always uphill.

If enough data is collected, a court of inquiry can prove anything.

Air strikes always overshoot the target, artillery always falls short.

When reviewing the radio frequencies that you just wrote down, the most important ones are always illegible.

To steal information from a person is called plagiarism. To steal information from the enemy is called gathering intelligence.

All-weather close air support doesn't work in bad weather.

The combat worth of a unit is inversely proportional to the smartness of its outfit and appearance.

Every command which can be misunderstood, will be.

Don't ever be the first, don't ever be the last and don't ever volunteer to do anything.

If your positions are firmly set and you are prepared to take the enemy assault on, he will bypass you.

If your ambush is properly set, the enemy won't walk into it.

The more stupid the leader is, the more important missions he is ordered to carry out.

The self-importance of a superior is inversely proportional to his position in the hierarchy.

Success occurs when no one is looking, failure occurs when the General is watching.

The enemy never monitors your radio frequency until you broadcast on an unsecured channel.

Never tell a superior officer that you have nothing to do.

If only one solution can be found for a field problem, then it is usually a stupid solution.

The quartermaster has only two sizes, too large and too small.

- Murphy's war laws

✧✧✧

A thing of orchestrated hell—a terrible symphony of light and flame.

- Edward R. Murrow, in a broadcast about his flight in a RAF Lancaster bombing Berlin. The famous broadcast became known as 'Orchestrated Hell.'
3 December 1943

✧✧✧

To master the virtue of the long sword is to govern the world and oneself, thus the long sword is the basis of strategy. The principle is "strategy by means of the long sword". If he attains the virtue of the long sword, one man can beat ten men. Just as one man can beat ten, so a hundred men can beat a thousand, and a thousand men can beat ten thousand. In my strategy (ARETE), one man is the same as ten thousand, so this strategy is the complete warrior's craft.

- Musashi Miyamoto

✧✧✧

If the enemy stays spirited it is difficult to crush him.

- Musashi Miyamoto

✧✧✧

If you consciously try to thwart opponents, you are already late.

- Miyamoto Mushashi (Philosopher/Samurai, 1645)

✧✧✧

In fighting and in everyday life you should be determined though calm. Meet the situation without tenseness yet not recklessly, your spirit settled yet unbiased. An elevated spirit is weak and a low spirit is weak. Do not let the enemy see your spirit.

- Miyamoto Mushashi (Philosopher/Samurai, 1645)

✧✧✧

Do nothing which is of no use.

- Miyamoto Mushashi (Philosopher/Samurai, 1645)

✧✧✧

You win battles by knowing the enemy's timing, and using a timing which the enemy does not expect.

- Miyamoto Mushashi (Philosopher/Samurai, 1645)

✧✧✧

Study strategy over the years and achieve the spirit of the warrior. Today is victory over yourself of yesterday; tomorrow is your victory over lesser men.

- Miyamoto Mushashi (Philosopher/Samurai, 1645)

✧✧✧

Perceive that which cannot be seen with the eye.

- Miyamoto Mushashi (Philosopher/Samurai, 1645)

✧✧✧

In battle, if you make your opponent flinch, you have already won.

- Miyamoto Musashi (The Book of Five Rings)

✧✧✧

You should not have any special fondness for a particular weapon, or anything else, for that matter. Too much is the same as not enough. Without imitating anyone else, you should have as much weaponry as suits you.

- Miyamoto Musashi (The Book of Five Rings)

✧✧✧

It is difficult to understand the universe if you only study one planet.

- Miyamoto Musashi (The Book of Five Rings)

✧✧✧

If you wish to control others you must first control yourself.

- Miyamoto Musashi (The Book of Five Rings)

✧✧✧

If you do not control the enemy, the enemy will control you.

- Miyamoto Musashi (The Book of Five Rings)

✧✧✧

To become the enemy, see yourself as the enemy of the enemy.

- Miyamoto Musashi (The Book of Five Rings)

✧✧✧

Think lightly of yourself and deeply of the world.

- Miyamoto Musashi (The Book of Five Rings)

✧✧✧

The important thing in strategy is to suppress the enemy's useful actions but allow his useless actions.

- Miyamoto Musashi (The Book of Five Rings)

✧✧✧

You can only fight the way you practice.

- Miyamoto Musashi (The Book of Five Rings)

✧✧✧

In all forms of strategy, it is necessary to maintain the combat stance in everyday life and to make your everyday stance your combat stance. You must research this well.

- Miyamoto Musashi

✧✧✧

Generally speaking, the way of the warrior is resolute acceptance of death.

- Miyamoto Musashi, 1645

✧✧✧

Was Islam spread by them through force and coercion? No. They preached Islam by personal example.

- General Pervez Musharraf

✧✧✧

There is nothing wrong with intellectual differences flowing from freedom of thought as long as such differences remain confined to intellectual debates.

- General Pervez Musharraf

✧✧✧

Most of the people in fact were against my writing this book at this moment, but like a good military leader, I took the decision against the major part of their advice.

- General Pervez Musharraf
at the launch of his book, 'In the Line of Fire".

✧✧✧

Considered purely in military terms, the Kargil operations were a landmark in the history of the Pakistani army.

- General Pervez Musharraf

✧✧✧

For my part I prefer fifty thousand rifles to fifty thousand votes.

- Benito Mussolini (1883–1945)

✧✧✧

It is better to live one day as a lion than 100 years as a sheep.

- Benito Mussolini

✧✧✧

N

In war, there are no unwounded soldiers.

- Jose Narosky

✧✧✧

We, the willing, led by the unknowing, are doing the impossible for the ungrateful. We have now done so much for so long with so little, we are now capable of doing anything with nothing.

- Navy Quote

✧✧✧

Why do we kill people who are killing people to show that killing people is wrong?

- Holly Near

✧✧✧

Too often we do things back to front. Because we have a Defence Force which can perform certain functions, we tend to construct a whole national security policy around those functions.

- R. O'Neill

✧✧✧

CLIFTON CHAPEL
This is the Chapel: here, my son,
Your father thought the thoughts of youth,
And heard the words that one by one
The touch of Life has turn'd to truth.
Here in a day that is not far,
You too may speak with noble ghosts
Of manhood and the vows of war
You made before the Lord of Hosts.
To set the cause above renown,
To love the game beyond the prize,
To honour, while you strike him down,
The foe that comes with fearless eyes;
To count the life of battle good,
And dear the land that gave you birth,
And dearer yet the brotherhood
That binds the brave of all the earth.

- Sir Henry Newbolt

✧✧✧

VITAÏ LAMPADA

There's a breathless hush in the Close to-night —
Ten to make and the match to win —
A bumping pitch and a blinding light,
An hour to play and the last man in.
And it's not for the sake of a ribboned coat,
Or the selfish hope of a season's fame,
But his Captain's hand on his shoulder smote
"Play up! play up! and play the game!"
The sand of the desert is sodden red, —
Red with the wreck of a square that broke; —
The Gatling's jammed and the colonel dead,
And the regiment blind with dust and smoke.
The river of death has brimmed his banks,
And England's far, and Honour a name,
But the voice of schoolboy rallies the ranks,
"Play up! play up! and play the game!"
This is the word that year by year
While in her place the School is set
Every one of her sons must hear,
And none that hears it dare forget.
This they all with a joyful mind
Bear through life like a torch in flame,
And falling fling to the host behind —
"Play up! play up! and play the game!"

- Sir Henry Newbolt

✧✧✧

No great political deed has been accomplished on the basis of a great vision of what the future would hold.

- Reinhold Niebuhr

✧✧✧

God, grant me the serenity to accept the things I cannot change,
Courage to change the things I can,
And wisdom to know the difference.

- Reinhold Niebuhr

✧✧✧

Beware that, when fighting monsters, you yourself do not become a monster...for when you gaze long into the abyss, the abyss gazes also into you.

- Friedrich Nietzsche

✧✧✧

Whatever doesn't kill you makes you stronger.

- Frederick Nietzsche

✧✧✧

No captain can do very wrong if he places his ship alongside that of an enemy.

- Admiral Horatio Nelson

✧✧✧

Gentlemen, when the enemy is committed to a mistake we must not interrupt him too soon.

- Admiral Horatio Nelson

✧✧✧

I have only one eye, I have a right to be blind sometimes... I really do not see the signal!

- Admiral Horatio Nelson

✧✧✧

Our country will, I believe, sooner forgive an officer for attacking an enemy than for letting it alone.

- Admiral Horatio Nelson

✧✧✧

Recollect that you must be a seaman to be an officer and also that you cannot be a good officer without being a gentleman.

- Admiral Horatio Nelson

✧✧✧

If a man consults whether he is to fight, when he has the power in his own hands, it is certain that his opinion is against fighting.

- Admiral Horatio Nelson

✧✧✧

Terrorism is carried out purposefully, in a cold-blooded, calculated fashion. The declared goals of the terrorist may change from place to place. He supposedly fights to remedy wrongs—social, religious, national, racial. But for all these problems his only solution is the demolition of the whole structure of society. No partial solution, not even the total redressing of the grievance he complains of, will satisfy him—until our social system is destroyed or delivered into his hands.

- Benjamin Netanyahu, International Terrorism

✧✧✧

When I say that terrorism is war against civilization, I may be met by the objection that terrorists are often idealists pursuing worthy ultimate aims—national or regional independence, and so forth. I do not accept this argument. I cannot agree that a terrorist can ever be an idealist, or that the objects sought can ever justify terrorism. The impact of terrorism, not merely on individual nations, but on humanity as a whole, is intrinsically evil, necessarily evil and wholly evil.

- Benjamin Netanyahu, International Terrorism

✧✧✧

It is the function of the Navy to carry the war to the enemy so that it will not be fought on U.S. soil.

- Admiral Chester W. Nimitz
Commander-in-Chief of the Pacific Fleet

✧✧✧

Among the men who fought on Iwo Jima, uncommon valour was a common virtue.

- Admiral Chester W. Nimitz (March 16, 1945)

✧✧✧

The best that science can devise and that naval organization can provide must be regarded only as an aid, and never as a substitute for good seamanship.

- Admiral Chester W. Nimitz

✧✧✧

I have just taken on a great responsibility. I will do my utmost to meet it.

- Admiral Chester W. Nimitz

✧✧✧

A ship is always referred to as "she" because it costs so much to keep one in paint and powder.

- Admiral Chester W. Nimitz

✧✧✧

That is not to say that we can relax our readiness to defend ourselves. Our armament must be adequate to the needs, but our faith is not primarily in these machines of defence but in ourselves.

- Admiral Chester W. Nimitz

✧✧✧

The basic objectives and principles of war do not change. The final objective in war is the destruction of the enemy's capacity and will to fight, and thereby force him to accept the imposition of the victor's will.

- Admiral Chester W. Nimitz

✧✧✧

Three favourite rules of thumb: Is the proposed operation likely to succeed? What might be the consequences of failure? Is it in the realm of practicality in terms of materiel and supplies?

- Admiral Chester W. Nimitz

✧✧✧

A riot is a spontaneous outburst. A war is subject to advance planning.

- President Richard M. Nixon

✧✧✧

A man is not finished when he is defeated. He is finished when he quits.

- President Richard M. Nixon

✧✧✧

If you want to make beautiful music, you must play the black and the white notes together.

- President Richard M. Nixon

✧✧✧

Only if you have been in the deepest valley, can you ever know how magnificent it is to be on the highest mountain.

- President Richard M. Nixon

✧✧✧

The Chinese use two brush strokes to write the word 'crisis.' One brush stroke stands for danger; the other for opportunity. In a crisis, be aware of the danger—but recognize the opportunity.

- President Richard M. Nixon

✧✧✧

If you take no risks, you will suffer no defeats. But if you take no risks, you win no victories.

- President Richard M. Nixon

✧✧✧

Never say no when a client asks for something, even if it is the moon. You can always try, and anyhow there is plenty of time afterwards to explain that it was not possible.

- President Richard M. Nixon

✧✧✧

Any change is resisted because bureaucrats have a vested interest in the chaos in which they exist.

- President Richard M. Nixon

✧✧✧

The bastards have never been bombed like they're going to be bombed this time.

- President Richard M. Nixon
during the 1972 North Vietnamese Easter Offensive

✧✧✧

Perhaps my dynamite plants will put an end to war sooner than your [pacifist] congresses. On the day two army corps can annihilate each other in one second all civilized nations will recoil from war in horror.

- Alfred Nobel

✧✧✧

Simply put, power is the ability to effect the outcomes you want, and if necessary, to change the behavior of others to make this happen.

- Joseph S. Nye Jr., 2002
The Paradox of American Power

✧✧✧

If we define *security* as the absence of acute threats to the minimal acceptable of the basic values that a people consider essential to its survival, then the economic dimension is important both as a potential instrument of threat to basic values and as one of the basic values itself.

- Joseph Nye

✧✧✧

Soft power is a country's cultural and ideological appeal. It is the ability to get desired outcomes through attraction instead of force. It works by convincing others that they should follow you or getting them to agree to norms and institutions that produce behaviour you want. Soft power depends largely on the persuasiveness of information. If a country can make its position attractive in the eyes of others and strengthen international institutions that encourage others to define their interests in compatible ways, it may not need to expend as many traditional economic or military resources. In today's global information age, soft power is becoming increasingly important.

- Joseph S. Nye Jr.

✧✧✧

The structure of power in the information age is like a three-dimensional chess game. On the top board, where the game is military issues, the U.S. is the sole superpower. On the middle board, where economics are played, the U.S., Europe and Japan account for nearly two-thirds of world product. On the bottom board of transnational global relations that cross boundaries outside the control of governments, power is widely dispersed among actors who range from bankers to terrorists.

- Joseph S. Nye Jr.

✧✧✧

Successful leadership may rest more upon soft power than in the past, but the prize will go to those with the contextual intelligence to manage the combination of soft and hard power into smart power.

- Joseph S. Nye Jr.

✧✧✧

O

In an interconnected world, the defeat of international terrorism—and most importantly, the prevention of these terrorist organizations from obtaining weapons of mass destruction—will require the cooperation of many nations. We must always reserve the right to strike unilaterally at terrorists wherever they may exist. But we should know that our success in doing so is enhanced by engaging our allies so that we receive the crucial diplomatic, military, intelligence, and financial support that can lighten our load and add legitimacy to our actions. This means talking to our friends and, at times, even our enemies.

- President Barack Obama,
20 November 2006

✧✧✧

It is the soldier, not the reporter, who gives us freedom of the press. It is the soldier, not the poet, who gives us the freedom of speech. It is the soldier, not the campus organizer, who gives us the freedom to demonstrate. It is the soldier, who salutes the flag, who serves under it, and whose coffin is draped by the flag, who allows the protestor to burn the flag.

- Father Davis Edward O' Brien Sr. USMC

✧✧✧

No commander who wishes to win ignores terrain or weather. No commander who wishes to win ignores the military capabilities of his enemies. Better commanders understand the confluence of terrain, weather, and enemy capabilities in formulating a course of action. But the best commanders in history used the cultural dimension as an integral part of warfare.

- Lieutenant Colonel Thomas P. Odom,
U.S. Army

✧✧✧

Anybody who doesn't have fear is an idiot. It's just that you must make the fear work for you. Hell, when somebody shot at me, it made me madder than hell, and all I wanted to do was shoot back.

- General Robin Olds, USAF

✧✧✧

The most important thing is to have a flexible approach. . . . The truth is no one knows exactly what air fighting will be like in the future. We can't say anything will stay as it is, but we also can't be certain the future will conform to particular theories, which so often, between the wars, have proved wrong.

- Brigadier General Robin Olds, USAF

✧✧✧

The best way to defend the bombers is to catch the enemy before it his in position to attack. Catch them when they are taking off, or when they are climbing, or when they are forming up. Don't think you can defend the bomber by circling around him. It's good for the bombers morale, and bad for tactics.

- Brigadier General Robin Olds, USAF

✧✧✧

What a man has not seen, he always expects will be greater than it really is.

- Onasander 1 A.D.

✧✧✧

If they (the young pilots) are on land, they would be bombed down, and if they are in the air, they would be shot down. That's sad...Too sad...To let the young men die beautifully, that's what *Tokko* is. To give beautiful death, that's called sympathy.

- Admiral Takijiro Onishi

(Note: Tokko means suicidal attack in Japanese)

✧✧✧

We knew the world could not be the same. A few people laughed, a few people cried. Most people were silent. I remembered the line from the Hindu scripture, the Bhagavad Gita: I am became Death, the destroyers of worlds. I suppose we all thought that, one way or another.

- Robert Oppenheimer

✧✧✧

The atomic bomb made the prospect of future war unendurable. It has led us up those last few steps to the mountain pass; and beyond there is a different country.

- Robert Oppenheimer

✧✧✧

We sleep safely in our beds because rough men stand ready in the night to visit violence on those who would do us harm.

- George Orwell

✧✧✧

He who controls the past controls the future, and he who controls the present controls the past.

- George Orwell (1984)

✧✧✧

The quickest way of ending a war is to lose it.

- George Orwell

✧✧✧

We were once told that the aeroplane had "abolished frontiers"; actually it is only since the aeroplane became a serious weapon that frontiers have become definitely impassable.

- George Orwell

'You and the Atomic Bomb', Tribune, London, 19 October 1945

✧✧✧

It is the act of a coward to wish for death.

- Ovid, 43 B.C.-18 A.D.

✧✧✧

Habits change into character.

- Ovid

✧✧✧

Resist beginnings; the prescription comes too late when the disease has gained strength by long delays.

- Ovid

✧✧✧

It is right to be taught, even by the enemy.

-Ovid

✧✧✧

The goal of the military revolution is providing dominance of information battlespace in a volume of hundreds of kilometres on each side and ten trillion wavelengths deep.

- Admiral William Owens

✧✧✧

P

Nothing concentrates the military mind so much as the discovery that you have walked into an ambush.

- Thomas Packenham,
British historian

✧✧✧

Good logistics is combat power.

- Lieutenant General William G. Pagonis,
US Army

✧✧✧

These are the times that try men's souls. The summer soldier and the sunshine patriot will, in this crisis, shrink from the service of their country; but he that stands it now, deserves the love and thanks of man and woman. Tyranny, like hell, is not easily conquered; yet we have this consolation with us, that the harder the conflict, the more glorious the triumph. What we obtain too cheap, we esteem too lightly: it is dearness only that gives everything its value.

- Thomas Paine,
The American Crisis

✧✧✧

It is impossible to calculate the moral mischief, if I may so express it, that mental lying has produced in society. When a man has so far corrupted and prostituted the chastity of his mind as to subscribe his professional belief to things he does not believe he has prepared himself for the commission of every other crime.

- Thomas Paine, The Age of Reason

✧✧✧

Iman, Taqwa, Jihad fi Sabilillah

Faith, Piety and Holy War in the Path of Allah

- Pakistan Army motto

✧✧✧

When an officer is granted a commission in the army, he does not merely "take a job." He embarks upon a profession which demands as much study and research as any other profession. When a doctor or a lawyer obtains a degree, and sets up a practice, he cannot meet with success unless he is constantly endeavouring to gain fresh knowledge of his profession. This is just as true of a soldier. The only difference is that, whereas the doctor or the lawyer is compelled to earn his living by virtue of professional skill, a commission in the army brings with it a certain degree of social and economic security. It makes it possible for an officer to struggle through his years of service with the barest minimum of effort.

- Lieut.Colonel D.K. Palit
The Essentials of Military Knowledge

✧✧✧

The company commander on the ground, attempting to fight his battle could usually observe orbiting in tiers above him his battalion commander, brigade commander, assistant division commander, division commander, and even his field force commander. With all that advice from the sky, it is easy to imagine how much individual initiative and control the company commander could exert on the ground—until nightfall sent the choppers to roost.

- Dave R Palmer
(Summons of the Trumpet: US-Vietnam in Perspective)

✧✧✧

The eternal difficulty in human affairs is to combine experience with youth; we demand too many qualifications and the man who does qualify is middle-aged. After twenty years of routine subordination, the chance is gone forever.

- William Parkinson

✧✧✧

Intuition is often crucial in combat, and survivors learn not to ignore it.

- Colonel F.F. Parry, US Marine Corps

✧✧✧

In the field of observation, chance favours the prepared mind.

- Louis Pasteur

✧✧✧

No bastard ever won a war by dying for his country. He won it by making the other poor dumb bastard die for his country.

- General George S. Patton

✧✧✧

If a man does his best, what else is there?

- General George S. Patton

✧✧✧

Accept the challenges so that you can feel the exhilaration of victory.

- General George S. Patton

✧✧✧

I complacently demanded the impossible, that I had dared extreme occasion and that I not taken council of my fears.

- General George S. Patton, War As I Knew It

✧✧✧

To be a good soldier, a man must have discipline, self-respect, and pride in his unit and his country, a high sense of duty and obligation to his comrades and to his superiors, and self-confidence born of demonstrated ability.

- General George S. Patton, War As I Knew It

✧✧✧

Self confidence, the greatest military virtue, results from the demonstrated ability derived from the acquisition of all the preceding qualities and from exercise in the use of weapons.

- General George S. Patton, War As I Knew It

✧✧✧

You cannot be disciplined in great things and undisciplined in small things. There is only one sort of discipline—perfect discipline. Discipline is based on:

Pride in the profession of arms.

Meticulous attention to detail.

Mutual Respect and Confidence.

Discipline can only be obtained when leaders are as imbued with the sense of their lawful obligations to their men and to their country that they cannot tolerate negligence.

- General George S. Patton

✧✧✧

Discipline can only be obtained when all the officers are imbued with the sense of their awful obligation to their men and to their country that they cannot tolerate negligence. Officers who fail to correct errors or to praise excellence are valueless in peace and dangerous misfits in war.

- General George S. Patton

✧✧✧

In case of doubt, attack.

- General George S. Patton

✧✧✧

The only way to win a war is to attack and keep on attacking and after you have done that keep attacking some more.

- General George S. Patton

✧✧✧

Unless you do your best, the day will come when, tired and hungry, you will halt just short of the goal you were ordered to reach, and by halting you will make useless the efforts and deaths of thousands.

- General George S. Patton

✧✧✧

If everyone is thinking alike, then somebody isn't thinking.

- General George S. Patton

✧✧✧

Success is how high you bounce when you hit bottom.

- General George S. Patton

✧✧✧

Lead me, follow me, or get out of my way.

- General George S. Patton

✧✧✧

Courage is fear holding on a minute longer.

- General George S. Patton

✧✧✧

The successful general makes plans to fit circumstances, but does not try to create circumstances to fit plans.

- General George S Patton

✧✧✧

The badge of rank which an officer wears on his uniform is really a symbol of servitude to his men.

- General George S Patton

✧✧✧

Pressure makes diamonds.

- General George S. Patton

✧✧✧

May God have mercy upon my enemies, because I won't.

- General George S. Patton

✧✧✧

If you can't get them to salute when they should salute and wear the clothes you tell them to wear, how are you going to get them to die for their country?

- General George S. Patton

✧✧✧

Never tell people how to do things. Tell them what to do and they will surprise you with their ingenuity.

- General George S. Patton

✧✧✧

If we take the generally accepted definition of bravery as a quality which knows no fear, I have never seen a brave man. All men are frightened. The courageous man is the man who forces himself, in spite of his fear, to carry on. Discipline, pride, self-respect, self-confidence and the love of glory are attributes which will make a man courageous, even when he is afraid.

- General George S. Patton

✧✧✧

Watch what people are cynical about, and one can often discover what they lack.

- General George S. Patton

✧✧✧

Do your damnedest in an ostentatious manner all the time.

- General George S. Patton

✧✧✧

Wars may be fought with weapons, but they are won by men.

- General George S. Patton

✧✧✧

I am a soldier, I fight where I am told, and I win where I fight.

- General George S. Patton

✧✧✧

Sure, we want to go home. We want this war over with. The quickest way to get it over with is to go get the bastards who started it. The quicker they are whipped, the quicker we can go home. The shortest way home is through Berlin and Tokyo. And when we get to Berlin, I am personally going to shoot that paper hanging son-of-a-bitch Hitler. Just like I'd shoot a snake!

- General George S. Patton,
addressing his troops before Operation Overlord, June 5, 1944

✧✧✧

Make your plans to fit the circumstances.

- General George S. Patton

✧✧✧

Moral courage is the most valuable and usually the most absent characteristic in men.

- General George S. Patton

✧✧✧

There's a great deal of talk about loyalty from the bottom to the top. Loyalty from the top down is even more necessary and is much less prevalent. One of the most frequently noted characteristics of great men who have remained great is loyalty to their subordinates.

- General George S. Patton

✧✧✧

You're never beaten until you admit it.

- General George S. Patton

✧✧✧

It is only by doing things others have not that one can advance.

- General George S. Patton

✧✧✧

A pint of sweat will save a gallon of blood.

- General George S. Patton

✧✧✧

All glory is fleeting.

- General George S. Patton

✧✧✧

A leader is a man who can adapt principles to circumstances.

- General George S. Patton

✧✧✧

Success demands a high level of logistical and organizational competence.

- General George S. Patton

✧✧✧

Perpetual peace is a futile dream.

- General George S. Patton

✧✧✧

It is foolish and wrong to mourn the men who died. Rather we should thank God that such men lived.

- General George S. Patton

✧✧✧

The test of success is not what you do when you are on top. Success is how high you bounce when you hit bottom.

- General George S. Patton

✧✧✧

If I win I can't be stopped! If I lose I shall be dead.

- General George S. Patton

✧✧✧

Never tell people how to do things. Tell them what to do and they will surprise you with their ingenuity.

- General George S. Patton

✧✧✧

Audacity, audacity, always audacity.

- General George S. Patton

✧✧✧

Magnificent! Compared to war all other forms of human endeavour shrink to insignificance. God help me, I do love it so!

- General George S. Patton

✧✧✧

Officers are responsible, not only for the conduct of their men in battle, but also for their health and contentment when not fighting. An officer must be the last man to take shelter from fire, and the first to move forward. Similarly, he must be the last man to look after his own comfort at the close of a march. He must see that his men are cared for. The officer must constantly interest himself in the rations of the men. He should know his men so well that any sign of sickness or nervous strain will be apparent to him, and he can take such action as may be necessary.

- General George S. Patton,
War as I Knew It

✧✧✧

A good plan executed today is better than a perfect plan executed at some indefinite point in the future.

- General George S. Patton

✧✧✧

There is only one tactical principle which is not subject to change. It is to use the means at hand to inflict the maximum amount of wound, death, and destruction on the enemy in the minimum amount of time.

- General George S. Patton

✧✧✧

Go forward until the last round is fired and the last drop of gas is expended...then go forward on foot!

- General George S. Patton

✧✧✧

It is the unconquerable nature of man and not the nature of the weapon he uses that ensures victory.

- General George S. Patton

✧✧✧

Fixed fortifications are monuments to the stupidity of man.

- General George S. Patton

✧✧✧

If you are going to win any battle, you have to do one thing. You have to make the mind run the body. Never let the body tell the mind what to do... the body is never tired if the mind is not tired.

- General George S. Patton

✧✧✧

Just drive down that road, until you get blown up.

- General George S. Patton, about reconnaissance troops

✧✧✧

You must do your damndest and win. Remember that is what you live for. Oh you must! You have got to do something! Never stop until you have gained the top or a grave.

- George S. Patton

✧✧✧

No man is fit to command another that cannot command himself.

- William Penn

✧✧✧

If we hadn't been buddies and comrades, we probably would have fled in panic. Each man depended on his buddy. . . . We depended on one another for our very survival, so we trusted one another. As comrades, we would never let each other down.

- Herb Peppard
The Lighthearted Soldier: A Canadian's Exploits with the Black Devils in WW II

✧✧✧

Freedom is the sure possession of those who have the courage to defend it.

- Pericles

✧✧✧

A competent leader can get efficient service from poor troops while . . .an incapable leader can demoralize the best of troops

- General John J. Pershing

✧✧✧

A censor is an expert in cutting remarks. A censor is a man who knows more than he thinks you ought to.

- Laurence J. Peter

✧✧✧

A man convinced against his will is not convinced.

- Laurence J. Peter

✧✧✧

There are two kinds of failures: those who thought and never did and those who did and never thought.

- Laurence J Peter

✧✧✧

A man doesn't know what he knows until he knows what he doesn't know.

- Laurence J. Peter

✧✧✧

An economist is an expert who will know tomorrow why the things he predicted yesterday didn't happen today.

- Laurence J. Peter

✧✧✧

As a matter of fact is an expression that precedes many an expression that isn't.

- Laurence J. Peter

✧✧✧

Bureaucracy defends the status quo long past the time when the quo has lost its status. There is too, a law governing voluntary labour: people are always available for work in the past tense.

- Laurence J. Peter

✧✧✧

Everyone rises to their level of incompetence.

- Laurence J. Peter

✧✧✧

Expert: a man who makes three correct guesses consecutively.

- Laurence J. Peter

✧✧✧

If you don't know where you're going, you will probably end up somewhere else.

- Laurence J. Peter

✧✧✧

Soldiers' bellies are not satisfied with empty promises and hopes.

- Peter the Great

✧✧✧

Where a goat can go, a man can go; where a man can go, he can drag a gun.

- Colonel William Phillips, 1777

✧✧✧

A force engaged is out of the hand of its commander.

- Colonel Charles Ardant du Picq
Battle Studies 1880

✧✧✧

The man is the first weapon of battle. Let us study the soldier for it is he who gives a reality to it.

- Colonel Ardant du Picq

✧✧✧

The instruments of battle are valuable only if one knows how to use them.

- Colonel Ardant du Picq

✧✧✧

In battle, two moral forces, even more than two material forces, are in conflict. The stronger conquers; the victor has often lost by fire more than the vanquished. Moral effect does not come entirely from the destructive power, real and effective as it may be. It comes, above all, from its presumed threatening power, present in the form of reserves threatening to renew the battle, of troops that appear on the flank, even of a determined frontal attack.

- Colonel Ardant du Picq

✧✧✧

Only study of the past can give us a sense of reality, and show us how the soldier will fight in the future.

- Colonel Ardant du Picq

✧✧✧

If one does not wish bonds broken, one should make them elastic and thereby strengthen them.

- Colonel Ardant du Picq

✧✧✧

Man does not enter battle to fight, but for victory. He does everything that he can to avoid the first and obtain the second.

- Colonel Ardant du Picq

✧✧✧

It often happens, that those who discuss war, taking the weapon for the starting point, assume unhesitatingly that the man called to serve it will always use it as contemplated and ordered. Man is flesh and blood; he is body and soul.

- Colonel Ardant du Picq

✧✧✧

A wise organization is that the personnel of combat groups changes as little as possible so that comrades in peacetime manoeuvres shall be comrades in war. From living together, and always obeying the same chiefs, from commanding the same men, from sharing fatigue and rest, from cooperation among men who quickly understand each other in the execution of war-like movements, may be bred brotherhood, professional knowledge, sentiment, and, above all, unity. Now confidence appears. It is intimate confidence, firm and conscious, which does not forget itself in the heat of battle and which alone makes true combatants.

- Colonel Ardant du Picq

✧✧✧

For me as a soldier, the smallest detail caught on the spot and in the heat of action is more instructive than all the Thiers and Jominis in all the world.

- Colonel Ardant du Picq, Battle Studies

✧✧✧

Absolute bravery, which does not refuse battle even on unequal terms, trusting only to God or to destiny, is not natural in man; it is the result of moral culture. ... Courage, that is the temporary domination of will over instinct, brings about victory.

- Colonel Ardant du Picq, Battle Studies

✧✧✧

Battle is the final objective of armies and man is the fundamental instrument in battle. Nothing can be wisely prescribed in an army...without exact knowledge of the fundamental instrument, man, and his state of mind, his morale, at the instant of combat.

- Colonel Ardant du Picq
Battle Studies: Ancient and Modern Battle

✧✧✧

The art of war is subjected to many modifications by industrial and scientific progress. But one thing does not change, the heart of man. In the last analysis, success in battle is a matter of morale. In all matters which pertain to an army, organization, discipline and tactics, the human heart in the supreme moment of battle is the basic factor. It is rarely taken into account; and often strange errors are the result.

- Colonel Ardant du Picq
Battle Studies; Ancient and Modern Battle

✧✧✧

Espirit de corps is secured in war. But war becomes shorter and shorter and more and more violent. Consequently secure esprit de corps in advance.

- Colonel Ardant du Picq

✧✧✧

Courage, that is the temporary domination of will over instinct, brings about victory.

- Colonel Ardant du Picq

✧✧✧

Note the army organizations and tactical formations on paper are always determined from the mechanical point of view neglecting the essential coefficient of morale. They are almost always wrong. When confidence is placed in superiority of material means, valuable as they are against an enemy at a distance, it may be betrayed by the actions of the enemy. If he closes with you in spite of your superiority in means of destruction, the morale of the enemy mounts with the loss of your confidence.

- Colonel Ardant du Picq

✧✧✧

Four brave men who do not know each other will not dare to attack a lion. Four less brave, but knowing each other well, sure of their reliability and consequently of mutual aid, will attack resolutely. There, is the science of the organization of armies in a nutshell.

- Colonel Ardant du Picq

✧✧✧

The instruments of battle are valuable only if one knows how to use them.

- Colonel Ardant du Picq

✧✧✧

Frederick liked to say that three men behind the enemy were worth fifty in front of him.

- Colonel Ardant du Picq

✧✧✧

Combat is the object, the cause of being, and the supreme manifestation of an army. Every measure that does not keep combat as the object of the army is fatal. All the resources accumulated in time of peace, all the training, and all the strategic calculations must have the goal of combat.

- Colonel Ardant du Picq

✧✧✧

Nothing can wisely be prescribed in any army... without exact knowledge of the fundamental instrument, man, and his state of mind, his morale, at the instant of combat.

- Colonel Ardant du Picq

✧✧✧

Generals of genius draw from the human heart ability to execute a surprising variety of movements which vary the routine; the mediocre ones, who have no eyes to read readily, are doomed to the worst errors.

- Colonel Ardant du Picq

✧✧✧

Courage, that is the temporary domination of will over instinct, brings about victory.

- Colonel Ardant du Picq

✧✧✧

Aviation is to serve only for reconnaissance and only in this direction should it be used.

- Marshal Josef Pilsudski, 1929

✧✧✧

The penalty good men pay for indifference to public affairs, is to be ruled by evil men.

- Greek philosopher Plato (429-347 B.C.)

✧✧✧

Only the dead have seen the end of war.

- Plato

✧✧✧

Nothing can be more important than that the work of a soldier should be well done.

- Plato

✧✧✧

The mind is not a vessel to be filled but a fire to be kindled.

- Plutarch

✧✧✧

I don't need a friend who changes when I change and who nods when I nod; my shadow does that much better.

- Plutarch

✧✧✧

Character is long-standing habit.

- Plutarch

✧✧✧

To make no mistakes is not in the power of man; but from their errors and mistakes the wise and good learn wisdom for the future.

- Plutarch

✧✧✧

The measure of a man is the way he bears up under misfortune.

- Plutarch

✧✧✧

The wildest colts make the best horses.

- Plutarch

✧✧✧

God is the brave man's hope, and not the coward's excuse.

- Plutarch

✧✧✧

Courage stands halfway between cowardice and rashness, one of which is a lack, the other an excess of courage.

- Plutarch

✧✧✧

Socrates thought that if all our misfortunes were laid in one common heap, whence every one must take an equal portion, most persons would be contented to take their own and depart.

- Plutarch

✧✧✧

So very difficult a matter is it to trace and find out the truth of anything by history.

- Plutarch

✧✧✧

A ship without Marines is like a garment without buttons.

- Admiral David D. Porter, US Navy

✧✧✧

It is the soldier, not the reporter, who has given us freedom of the press. It is the soldier, not the poet, who has given us freedom of speech. It is the soldier, not the organizer, who has given us the freedom to demonstrate. It is the soldier, who salutes the flag, who serves beneath the flag, and whose coffin is draped by the flag, who allows the protestor to burn the flag.

- Charles M. Province

✧✧✧

The future starts today, not tomorrow.

- Pope John Paul II

✧✧✧

Courage is as often the outcome of despair as of hope; in the one case we have nothing to lose, in the other everything to gain.

- Diane De Poitiers

✧✧✧

I felt cold, muddy and dirty.... [But] you keep going because you feel it's your duty to keep going. You don't think about it. Your men are going on because *you* are.

- Major George Pearkes
awarded the Victoria Cross in October 1917.
Leading a company of Canadian infantry in an attack on the defences of Passchendaele, he was wounded in the thigh by a German counter-barrage. He briefly pondered whether he would go back to seek first aid. But, when he saw his men hesitating in their advance, the training that had been instilled in him when he was young in a public school in England came back to him.

✧✧✧

Part of the happiness of life consists not in fighting battles, but in avoiding them. A masterly retreat is in itself a victory.

- Norman Vincent Peale

✧✧✧

Speak properly, and in as few words as you can, but always plainly; for the end of speech is not ostentation, but to be understood.

- William Penn

✧✧✧

I expect to pass through life but once. If therefore, there be any kindness I can show, or any good thing I can do to any fellow being, let me do it now, and not defer or neglect it, as I shall not pass this way again.

- William Penn

✧✧✧

Governments, like clocks, go from the motion men give them, and as governments are made and moved by men, so by them they are ruined too. Wherefore governments rather depend upon men than men upon governments. Let men be good, and the government cannot be bad; if it be ill, they will cure it. But, if men be bad, let the government be ever so good, they will endeavour to warp and spoil it to their turn.

- William Penn

✧✧✧

Be reserved, but not sour
Bold, but not rash
Humble, but not servile
Patient, but not insensible
Constant, but not obstinate
Cheerful, not light
Rather sweet than familiar,
Familiar than intimate
And intimate with very few
And upon very few grounds.

- William Penn

✧✧✧

No man is fit to command another that cannot command himself.

- William Penn

✧✧✧

We have met the enemy and they are ours ...

- Oliver Hazard Perry
Dispatch to Major General William Henry Harrison after the Battle of Lake Erie, 10 September 1813

✧✧✧

We must not allow ourselves to be driven out by terrorists. That would not only reward and encourage terrorism; it would jeopardize our ability to defend our vital national interests.

- Defence Secretary William J. Perry, 9 July 1996

✧✧✧

In the moment of action remember the value of silence and order.

- Phormio of Athens

✧✧✧

Another popular fallacy is to suppose that flying machines could be used to drop dynamite on an enemy in time of war.

- William H. Pickering, Aeronautics, 1908.

✧✧✧

I don't care how they dress so long as they mind their fighting.

- Sir Thomas Picton, British general killed at Waterloo, 1815, when questioned about the slovenly appearance of his troops

✧✧✧

I am surrounded by fearful odds that will overcome me and my valiant men; but I am pleased to die fighting for my beloved country.

- General Gregorio H. Del Pilar (1875–1899), hero of the Philippine-American war

✧✧✧

The victory of a war is not determined by the size of an army or the quality of armaments. Factors like unshakable determination, heroism, and desire for liberation determine victory.

- Vellupillai Pirabhakaran
Sri Lanka LTTE leader

✧✧✧

If I were an American, as I am an Englishman, while a foreign troop was landed in my country I never would lay down my arms, Never! Never! Never!

- William Pitt, 1708-1778

✧✧✧

In these matters the only certainty is that nothing is certain.

- Pliny the Elder

✧✧✧

It is generally much more shameful to lose a good reputation than never to have acquired it.

- Pliny the Elder

✧✧✧

No mortal man, moreover is wise at all moments.

- Pliny the Elder

✧✧✧

The best plan is to profit by the folly of others.

- Pliny the Elder

✧✧✧

The depth of darkness to which you can descend and still live is an exact measure of the height to which you can aspire to reach.

- Pliny the Elder

✧✧✧

An object in possession seldom retains the same charm that it had in pursuit.

- Pliny the Younger

✧✧✧

Gentlemen, I don't know whether we will make history tomorrow, but we will certainly change geography.

- Sir Herbert Plumer
(At a press conference the day before the blowing up of Messines Ridge, 6 June 1917)

✧✧✧

The most instructive, indeed the only method of learning to bear with dignity the vicissitude of fortune, is to recall the catastrophes of others.

- Polybius

✧✧✧

A ship without Marines is like a garment without buttons.

- Admiral David D. Porter, USN

✧✧✧

The day soldiers stop bringing you their problems is the day you have stopped leading them. They have either lost confidence that you can help them or concluded that you do not care. Either case is a failure of leadership.

- General Colin Powell

✧✧✧

Do not be bewildered by experts. They often possess more data than judgement.

- General Colin Powell

✧✧✧

Don't let adverse facts stand in the way of a good decision.

- General Colin Powell

✧✧✧

The mission is primary, followed by taking care of your soldiers.

- General Colin Powell

✧✧✧

- It ain't as bad as you think. It will look better in the morning.
- Get mad, then get over it.
- Avoid having your ego so close to your position that when your position falls, your ego goes with it.
- Be careful what you choose. You may get it.
- Do not let adverse facts get in the way of a good decision.
- You cannot make someone else's choices. You should not let someone else make yours.
- Check small things.
- Share credit.
- Remain calm. Be kind.
- Have a vision. Be demanding.
- Do not take counsel of your fears or naysayers.
- Perpetual optimism is a force-multiplier.

- General Colin Powell (Colin Powell's rules)

✧✧✧

Over the years, the United States has sent many of its fine young men and women into great peril to fight for freedom beyond our borders. The only amount of land we have ever asked for in return is enough to bury those that did not return.

- General Colin Powell

✧✧✧

Well, we did not build those bombers to carry crushed rose petals.

- Gen Thomas S. Power, USAF

✧✧✧

Give a man a fire, and he's war for a day. But set fire to him, and he's warm for the rest of his life.

- Terry Pratchett

✧✧✧

A visitor from Mars could easily pick out the civilized nations. They have the best implements of war.

- Herbert V. Prochnow

✧✧✧

The real voyage of discovery consists not of finding new lands, but of seeing the territory with new eyes.

- Marcel Proust

✧✧✧

It is the soldier, not the reporter, who has given us freedom of the press. It is the soldier, not the poet, who has given us freedom of speech. It is the soldier, not the organizer, who has given us the freedom to demonstrate. It is the soldier, who salutes the flag, who serves beneath the flag, and whose coffin is draped by the flag, who allows the protestor to burn the flag.

- Charles M. Province

✧✧✧

Paper-work will ruin any military force.

- Lieutenant-General Lewis B. "Chesty" Puller
US Marine Corps

✧✧✧

They are in front of us, behind us, and we are flanked on both sides by an enemy that outnumbers us 29:1. They can't get away from us now!

- Lieutenant-General Lewis B. "Chesty" Puller

✧✧✧

The concentration of troops can be done fast and easy, on paper.

- Radomir Putnik, Serbian field marshal, 1912-17

✧✧✧

Q

Our Gods and soldiers we alike adore,
Even at the brink of danger, not before.
After deliverance, both alike required,
Our Gods forgotten and our soldiers slighted

- Francis Quarles

✧✧✧

So long as peace is not attained by law (so argue the advocates of armaments) the military protection of a country must not be undermined, and until such is the case disarmament is impossible.

- Ludwig Quidde

✧✧✧

When distrust exists between governments, when there is a danger of war, they will not be willing to disarm even when logic indicates that disarmament would not affect military security at all.

- Ludwig Quidde

✧✧✧

A strategy is the pattern or plan that integrates an organization's major goals, policies, and action sequences into a cohesive whole. A well-formulated strategy helps to marshal and allocate an organization's resources into a unique and viable posture based upon its relative internal competencies and shortcomings, anticipated changes in the environment, and contingent moves by intelligent opponents.

- James Brian Quinn,
Strategies for Change: Logical Incrementalism

✧✧✧

R

... the basic principles of the military services are unchangeable. Courage and candour, obedience and comradeship, love of fatherland and loyalty to the State: these are ever the distinguishing characteristics of the soldier and sailor. Building character through intelligent training and education is always the first and greatest goal.

- Grand Admiral Erich Raeder,
Commander in Chief of the German Navy

◊◊◊

As I prepare for this last mission, I am a bit homesick...Mother and Dad, you are very close to me, and I long so to talk to you. America has asked much of our generation, but I'm glad to give her all I have because she has given me so much

- Sergeant Arnold Rahe U.S. Army Air Forces, WWII
Killed in Action over France... from a letter to his parents

◊◊◊

For whosoever commands the sea commands the trade; whosoever commands the trade of the world commands the riches of the world, and consequently the world itself.

- Sir Walter Raleigh

◊◊◊

Trust few men; above all, keep your follies to yourself.

- Sir Walter Raleigh

◊◊◊

Better it were not to live than to live a coward.

- Sir Walter Raleigh

◊◊◊

Prevention is the daughter of intelligence.

- Sir Walter Raleigh

◊◊◊

The cavalry, in particular, were not friendly to the aeroplane, which it was believed, would frighten the horses.

- Sir Walter Alexander Raleigh
(This is not the Sir Walter Raleigh who was beheaded nearly three hundred years earlier.)
This Sir Walter became the official historian of the RAF

◊◊◊

He spoke, "My name is Clifford Rafuse." Then, taking his swagger stick and touching the crown insignia on his arm, he would say, "I am a Sergeant-Major. You will not address me as Clifford, Cliff, Rafuse, sir, hey you, or any of the foul names you really think of me in your pea-sized brains. I am a Sergeant-Major—here, in the shower, in the latrine, in my drawers, in my pyjamas, or when I am dead. I am, and always will be Sergeant-Major Rafuse to you. If you pumpkin-heads see me on the street twenty-five years from now—and most of you won't survive this training to live that long- I will still be addressed as Sergeant-Major by you. Do you understand that?"

Then bellowing again, he demanded they scream an answer:
"Yes, Sergeant-Major."

He would then go on: "I'm not your mother; I won't tuck you in bed; and I won't be your pal. I will make you bleedin', sloppy, unwashed, useless, pudgy loafers who thought this army was a holiday camp into battle shape. I shall turn your pudgy asses into such shape that you will have muscles in your defecation. Some few of you who fooled your way through some little school may think you are smart and will think you will fool me because you know the ABC's! You will not fool me; you are not smart. And when I say "jump", you say "How high". When I say "defecate", you say "Yes, Sir, and what colour, Sir?. I shall make you baggy, civilian lot of unwashed, sloppy, buggers into cleaned, shined, well-spoken, and obedient battle-ready troops. Or.. you will suffer a fate and terror worse than heck.

"Your Mother can't save you. Nobody is tougher than I am. I am tougher than any Kraut you ever encounter. Even the Padre is scared of me. I'll march you, drill you, train you, punish you, and toughen you into soldiers. Don't talk back; don't complain, even to the Padre; because my words will even bring tears to his eyes. Now tighten up those soft pudgy asses, pull in those sagging chins, and suck in those baggy guts. Hands by your sides with thumbs down the seams of those potato-bag looking trousers.

"Like this," as he demonstrated, "and when you get that right, we'll take you ladies to a lovely King's breakfast of such quality you'll be glad when we let you work in our kitchen. Our next present to you slobs will be a visit to His Majesty's barber so as you can get that bleedin', mangled, lady-length, dirty, bug-infested civilian hairdo cut off. You will then, at least, not look like a bleedin' civilian, with a filthy mat on your head. Now fall out, ladies, and form up for the cookhouse. MARCH—quickly, before I lose my f...n' temper."

- Sergeant-Major Rafuse'
standard welcome to new recruits, Camp Aldershot, 1939.

✧✧✧

A long-range military problem is comparable to the problem of an owner of a racing stable who wants to win a horse race, to be run many years hence, on a track not yet built, between horses not yet born. To make matters worse, the possibility exists that, when the race is finally run, the rules may have changed, the track length altered and the horses replaced by greyhounds.

- RAND Corporation

⟡⟡⟡

The surest way to prevent a war is not to fear it.

- John Randolf, to the House of Representatives, 5 March 1808

⟡⟡⟡

You can no more win a war than you can win an earthquake.

- Jeannette Rankin, First Woman Member of Congress

⟡⟡⟡

The person who knows how will always have a job. The person who knows why will always be his boss.

- Diane Ravitch, speech, 1985

⟡⟡⟡

Fetishism for battle drills has been largely responsible for sanitizing imagination, creativity and mental mobility in infantry ranks. Battle drills are ... a set of reactions ... Conversely, tactics are a thought out plan to overcome the threat, the two are therefore dissimilar.

- Col Arjun Ray, quoted in the RUSI Journal, Autumn 1989

⟡⟡⟡

The greatest intensification of the horrors of war is a direct result of the democratisation of the State. So long as the army was a professional unit, the specialist function of a limited number of men, war remained a relatively harmless contest for power. But once it became everyman's duty to defend his home (or his political rights) warfare was free to range wherever that home might be, and to attack every form of life and property associated with that home.

- Herbert Read

⟡⟡⟡

We should declare war on North Vietnam. . . .We could pave the whole country and put parking strips on it, and still be home by Christmas.

- Ronald Reagan, 1965
40th President of the United States (1981–1989)

⟡⟡⟡

Some people live an entire lifetime and wonder if they have ever made a difference in the world, but the Marines don't have that problem.

- Ronald Reagan

⟡⟡⟡

Today we did what we had to do. They counted on America to be passive. They counted wrong.

- Ronald Reagan
In response to terrorism by Libya (1986)

✧✧✧

A people free to choose will always choose peace.

- Ronald Reagan

✧✧✧

Inspiration springs from great tradition...guard the traditions of your service, built in the foothills of the Rockies and in the air over Ploesti, MIG Alley, Red River Valley, and a thousand other places.

- President Ronald W. Reagan

✧✧✧

Of the four wars in my lifetime none came about because the United States was too strong.

- Ronald Reagan

✧✧✧

Before I refuse to take your questions, I have an opening statement.

- Ronald Reagan

✧✧✧

Government is like a baby. An alimentary canal with a big appetite at one end and no sense of responsibility at the other.

- Ronald Reagan

✧✧✧

Status quo, you know, is Latin for 'the mess we're in'.

- Ronald Reagan

✧✧✧

The most terrifying words in the English language are: I'm from the government and I'm here to help.

- Ronald Reagan

✧✧✧

Information is the oxygen of the modern age. It seeps through the walls topped by barbed wire, it wafts across the electrified borders.

- Ronald Reagan

✧✧✧

I couldn't help but say to [Mr. Gorbachev], just think how easy his task and mine might be in these meetings that we held if suddenly there was a threat to this world from another planet. [We'd] find out once and for all that we really are all human beings here on this earth together.

- Ronald Reagan, 1985

✧✧✧

We should declare war on North Vietnam. . . .We could pave the whole country and put parking strips on it, and still be home by Christmas.

- Ronald Reagan, 1965

✧✧✧

No arsenal or no weapon in the arsenals of the world is so formidable as the will and moral courage of free men and women.

- Ronald Reagan, First Inaugural Address, 20 January 1981

✧✧✧

Colonel Moore, I said, tell me something about leadership. I had hit a sensitive spot. He forged ahead.

Leadership! The greatest problem here is the leaders, and you have to find some way to weed out the weak ones. It's tough to do this when you're in combat. The platoon leaders who cannot command, who cannot foresee things, and who cannot act on the spur of the moment in an emergency are a distinct detriment.

It is hot here, as you can see. Men struggle; they get heat exhaustion. They come out vomiting, and throwing away equipment. The leaders must be leaders and they must be alert to establish straggler lines and stop this thing.

The men have been taught to take salt tablets, but the leaders don't see to this. Result, heat exhaustion. The good leaders seem to get killed; the poor leaders get the men killed. The big problem is leadership and getting the shoulder straps on the right people.

Sixty-millimeter Japanese mortar shells fell about thirty yards away and attacked a number of coconut trees. I lost interest in taking dictation and the colonel stopped talking. When the salvo was over and things were quiet again, Bryant Moore said, Where was I? You saw that patrol. I tell you this, not one man in fifty can lead a patrol in this jungle. If you can find out who the good patrol leaders are before you hit the combat zone, you have found out something.

I have had to get rid of about twenty-five officers because they just weren't leaders. I had to *make* the battalion commander weed out the poor junior leaders! This process is continuous. Our junior leaders are finding out that they must know more about their men. The good leaders know their men.

- Colonel Red Reeder
Account of his conversation with Colonel Bryant Moore on Guadalcanal in the book—Born at Reveille

✧✧✧

The more you sweat in peace, the less you bleed in war.

- Hyman G. Rickover 1986

✧✧✧

Find the enemy and shoot him down. Anything else is nonsense.

- Captain Manfred von Richthofen (The Red Baron), 1917

✧✧✧

There had been an incident on the march that could have had tragic results. Two British officers had seen Private A.W. Belyea of D Company [of the Second (Special Service) Battalion, The Royal Canadian Regiment] grab a stray chicken that crossed his path. Looting was anathema to the army, and Belyea was court-martialled. To set an example for the troops the brigade was formed up in a hollow square to hear the verdict. For poor Belyea the ordeal was terrifying as he stood alone, head bowed, awaiting the decision of the court. The verdict was hardly in doubt, and the offence could draw the death penalty. The officers who made up the court realized the maximum punishment did not fit the crime. Belyea was confined to barracks for 56 days, a meaningless punishment on the veldt. (From a related footnote—...Capt S.M. Rogers, who commanded D Company, told his men, "Now listen, boys, it wasn't for stealing the chicken that [Belyea] was going to be hung, it was for getting caught at it, so watch yourself.")

- Brian A. Reid, Our Little Army in the Field; The Canadians in South Africa 1899–1902, 1996

✧✧✧

As well, disease continued to take its deadly toll, including Private Smith from the CMR under the most unusual circumstances. He had fallen ill on the march and reported to the ambulance, only to find that the doctor was absent. The medical supplies were carried in a large wicker chest. On the inside of the lid was a printed sheet with a list of diseases, followed by a diagnosis for each as well as a remedy which had a number. Thus, if a soldier was constipated, he should take a number nine pill. Smith read the sheet carefully and decided he needed a number seven pill. Alas, the number seven bottle was empty. Instead, he decided to take a number two and a number five. In the words of Private Griesbach, "Mathematically he seemed to be deadright, but ... we buried him that night".

- Brian A. Reid, Our Little Army in the Field; The Canadians in South Africa 1899-1902, 1996

✧✧✧

Romans not only easily conquered those who fought by cutting, but mocked them too. For the cut, even delivered with force, frequently does not kill, when the vital parts are protected by equipment and bone. On the contrary, a point brought to bear is fatal at two inches; for it is necessary that whatever vital parts it penetrates, it is immersed. Next, when a cut is delivered, the right arm and flank are exposed. However, the point is delivered with the cover of the body and wounds the enemy before he sees it. ... However, they are given that double-weight shield frame and foil, so that when the recruit takes up real, lighter weapons, as if freed from the heavier weight, he will fight in greater safety and faster. But when field training was ended through negligence and laxity, the equipment (which the soldiers seldom put on) began to be seen as heavy.

- Flavius Vegetius Renatus (in Epitoma Rei Militari)

✧✧✧

Qui desiderat pacem, praeparet bellum.

[Who desires peace should prepare for war.]

- (Flavius) Vegetius (Renatus), De Rei Militari III, c. 375 AD

✧✧✧

It is necessary to know the sentiments of the soldiers on the day of an engagement. Their confidence or apprehensions are easily discovered by their looks, their words, their actions and their motions. No great dependence is to be placed on the eagerness of young soldiers for action, for the prospect of fighting is attractive to those who are strangers to it. On the other hand, it would be wrong to hazard an engagement, if old experienced soldiers testify to a disinclination to fight. A general, however, may encourage and animate his troops by proper exhortations and orations, especially if by his prophecies of a favourable result of the approaching action he can persuade them into the belief of an easy victory.

- Flavius Vegetius Renatus, The Military Institutions of the Romans

✧✧✧

Courage consists, not in blindly overlooking danger but in seeing and conquering it.

- Richter

✧✧✧

Courage is doing what you're afraid to do. There can be no courage unless you're scared.

- Captain Edward V. 'Eddie' Rickenbacker
Leading American Ace of WWI

✧✧✧

Military history, accompanied by sound criticism, is indeed the true school of war.

- Gen Matthew B. Ridgway

✧✧✧

The art of war teaches us to rely not on the likelihood of the enemy not coming, but on our own readiness to receive him; not on the chance of his not attacking, but rather on the fact that we have made our position unassailable.

- Field Marshall Roberts Bt, VC (1832–1914)

✧✧✧

I can speak from my own experience of what use a regimental band is to a regiment. I have seen men weary, worn out with fatigue, hot and smothered with dust, brighten up the moment they heard the tap of the drum, indicating that the band was going to play a lively quick step.

- Field-Marshal Lord Roberts, V.C.

✧✧✧

I feel sure I am right when I say that the less the Afghans see of us, the less they will dislike us.

- Field-Marshal Lord Roberts, V.C.

✧✧✧

The principles of war are the same as those of a siege. Fire must be concentrated on one point and as soon as the breach is made, the equilibrium is broken and the rest is nothing.

- Major General C.W. Robinson, Wars of the 19th Century

✧✧✧

Diplomats are just as essential to starting a war as soldiers are for finishing it.... You take diplomacy out of war, and the thing would fall flat in a week.

- Will Rogers

✧✧✧

You can't say that civilization don't advance, for in every war they kill you a new way.

- Will Rogers

✧✧✧

Success is neither magical or mysterious. Success is the natural consequence of consistently applying the fundamentals.

- Jim Rohn

✧✧✧

I have to follow them. I am their leader.

- Alexandre-Auguste Ledru-Rollin,
Leader of the French Revolution of 1848

✧✧✧

In the absence of orders, go find something and kill it.

- Field Marshal Erwin Rommel

✧✧✧

The best results are obtained from a commander whose ideas develop freely from the conditions around him and have not previously been channelized into any fixed pattern.

- Field Marshal Erwin Rommel

✧✧✧

The best form of welfare is first class training.

- Field Marshal Erwin Rommel

✧✧✧

There is no ideal solution to military problems; every course has its advantages. One must select that which seems best from the most varied aspects and then pursue it resolutely and accept the consequences.

- Field Marshal Erwin Rommel

✧✧✧

The future battle on the ground will be preceded by battle in the air. This will determine which of the contestants has to suffer operational and tactical disadvantages and be forced throughout the battle into adopting compromise solutions.

- General Erwin Rommel

✧✧✧

Be an example to your men, in your duty and in private life. Never spare yourself, and let the troops see that you don't in your endurance of fatigue and privation. Always be tactful and well-mannered and teach your subordinates to do the same. Avoid excessive sharpness or harshness of voice, which usually indicates the man who has shortcomings of his own to hide.

- Field Marshal Erwin Rommel

✧✧✧

One must not judge everyone in the world by his qualities as a soldier: otherwise we should have no civilization.

- Field Marshal Erwin Rommel

✧✧✧

I would rather he had given me one more division.

- Rommel, when Hitler made him a Field Marshal

✧✧✧

I have found again and again that in encounter action, the day goes to the side that is the first to plaster its opponents with fire. The man who lies low and awaits developments usually comes off second-best.

- Field Marshal Erwin Rommel

✧✧✧

It is fundamentally wrong simply to halt and look for cover without opening fire, or to wait for more forces to come up and take part in the action.

- Field Marshal Erwin Rommel

✧✧✧

The sole criterion for a commander in carrying out a given operation must be the time he is allowed for it, and he must use all his powers of execution to fulfil that task within that time.

- Field Marshal Erwin Rommel

✧✧✧

A commander's drive and energy often count for more than his intellectual powers—a fact that is not generally understood by academic soldiers, although for the practical man it is self-evident.

- Field Marshal Erwin Rommel

✧✧✧

One is forced again and again to re-learn the fact that standards set by precedent are based on something less than average performance, and for that reason, one should not submit to them.

- Field Marshal Erwin Rommel

✧✧✧

Anyone who has to fight, even with the most modern weapons, against an enemy in complete command of the air, fights like a savage against modern European troops, under the same handicaps and with the same chances of success.

- Field Marshal Erwin Rommel, Rommel Papers, 1953.

✧✧✧

But courage which goes against military expediency is stupidity, or, if it is insisted upon by a commander, irresponsibility.

- Erwin Rommel

✧✧✧

Don't fight a battle if you don't gain anything by winning.

- Field Marshal Erwin Rommel

✧✧✧

In a man-to-man fight, the winner is he who has one more round in his magazine.

- Field Marshal Erwin Rommel

✧✧✧

Sweat saves blood.

- Field Marshal Erwin Rommel

✧✧✧

The future battle on the ground will be preceded by battle in the air. This will determine which of the contestants has to suffer operational and tactical disadvantages and be forced throughout the battle into adoption compromise solutions.

- Field Marshal Erwin Rommel

✧✧✧

I'm still young, give me time.

- Field Marshal Gerd von Runstedt
In reply to a statement by Major General Gunther Blumentritt:
Everyone knows you have never lost a battle.

✧✧✧

Many people would sooner die than think; in fact, they do so.

- Bertrand Russell

✧✧✧

No commander should consider his appointment gazetted until it has been ratified in the hearts and minds of his men.

- Lt Gen Sir Dudley Russell

✧✧✧

The hammer shatters glass but forges steel.

- Russian proverb

✧✧✧

I am the Unknown Soldier
And maybe I died in vain,
But if I were alive and my country called,
I'd do it all over again.

- Billy Rose

✧✧✧

The universe is so vast and so ageless that the life of one man can only be justified by the measure of his sacrifice.

- Pilot Officer V.A. Rosewarne, 1940

✧✧✧

We will have to want peace, want it enough to pay for it, before it becomes an accepted rule.

- Eleanor Roosevelt, 1884-1964

✧✧✧

You gain strength, courage, and confidence by every experience in which you really stop to look fear in the face. You are able to say to yourself, 'I have lived through this horror. I can take the next thing that comes along.' You must do the thing you think you cannot do.

- Eleanor Roosevelt

✧✧✧

Nobody can make you feel inferior without your consent.

- Eleanor Roosevelt

✧✧✧

The only thing we have to fear is fear itself.

- Franklin D. Roosevelt

✧✧✧

Hitler built a fortress around Europe, but he forgot to put a roof on it.

- Franklin D. Roosevelt

✧✧✧

When you get to the end of your rope, tie a knot and hang on.

- Franklin D. Roosevelt

✧✧✧

It is not the critic who counts, not the man who points out how the strong man stumbled, or where the doer of deeds could have done better. The credit belongs to the man who is actually in the arena; whose face is marred by dust and sweat and blood; who strives valiantly; who errs and comes short again and again, because there is no effort without error or shortcoming, but who knows the great enthusiasms, the great devotions and spends himself in a worthy cause, who at the best, knows in the end the triumph of high achievement; and who, at the worse, if he fails, at least fails while daring greatly, so that his place shall never be with those cold and timid souls who know neither victory nor defeat.

- Theodore Roosevelt.

✧✧✧

Only those are fit to live who do not fear to die; and none are fit to die who have shrunk from the joy of life and the duty of life. Both life and death are parts of the same Great Adventure.

- Theodore Roosevelt

✧✧✧

Far better it is to dare mighty things, to win glorious triumphs, even though chequered by failure, than to take rank with those poor spirits who neither enjoy nor suffer much, because they live in the gray twilight that knows neither victory nor defeat.

-Theodore Roosevelt

✧✧✧

Whenever you are asked if you can do a job, tell 'em, "Certainly, I can!" Then get busy and find out how to do it.

-Theodore Roosevelt

✧✧✧

Let us speak courteously, deal fairly, and keep ourselves armed and ready.

- Theodore Roosevelt

✧✧✧

It is not the critic who counts; not the man who points out how the strong man stumbled, or where the doer of deeds could have done better. The credit belong to the man who is actually in the arena; whose face is marred by dust and sweat and blood; who strives valiantly; who errs and comes short again and again. Who knows the great enthusiasms, the great devotions, and spends himself in a worthy cause. Who at the best knows in the end the triumph of high achievement; and who at the worst, if he fails, at least fails while daring greatly. So that his place shall never be with those cold and timid souls who know neither victory nor defeat.

- Theodore Roosevelt

✧✧✧

A just war is in the long run far better for a man's soul than the most prosperous peace.

- President Theodore Roosevelt

✧✧✧

Such an experiment without actual conditions of war to support it is a foolish waste of time. . . . I once saw a man kill a lion with a 30-30 calibre rifle under certain conditions, but that doesn't mean that a 30-30 rifle is a lion gun.

- Theodore Roosevelt, Jr., U.S. Assistant Secretary of the Navy, War, regards Billy Mitchell's experimental sinking of the captured German battleship Ostfreisland. 1921.

✧✧✧

No man is worth his salt who is not ready at all times to risk his body, to risk his wellbeing, to risk his life, in a great cause.

- Theodore Roosevelt

✧✧✧

It is not the critic who counts, not the man who points out how the strong man stumbled, or where the doer of deeds could have done better. The credit belongs to the man who is actually in the arena; whose face is marred by the dust and sweat and blood; who strives valiantly; who errs and comes short again and again; who knows the great enthusiasms, the great devotions and spends himself in a worthy course; who at the best, knows in the end the triumph of high achievement, and who, at worst, if he fails, at least fails while daring greatly; so that his place shall never be with those cold and timid souls who know neither victory or defeat.

- Theodore Roosevelt

✧✧✧

Do not hit at all if it can be avoided, but never hit softly.

- Theodore Roosevelt

✧✧✧

Far better it is to dare mighty things, to win glorious triumphs, even though chequered by failure, than to take rank with those poor spirits who neither enjoy much nor suffer much, because they live in the gray twilight that knows not victory nor defeat.

- Theodore Roosevelt, Speech 10 April 1899

✧✧✧

No man is justified in doing evil on the ground of expediency.

- Theodore Roosevelt
The Strenuous Life: Essays and Addresses

✧✧✧

Speak softly and carry a big stick; you will go far.

- Theodore Roosevelt

✧✧✧

A man who is good enough to shed his blood for the country is good enough to be given a square deal afterwards.

- Theodore Roosevelt

✧✧✧

A vote is like a rifle; its usefulness depends upon the character of the user.

- Theodore Roosevelt

✧✧✧

Do what you can, with what you have, where you are.

- Theodore Roosevelt

✧✧✧

Don't hit at all if it is honourably possible to avoid hitting; but never hit soft.

- Theodore Roosevelt

✧✧✧

What do you want to achieve or avoid? The answers to this question are objectives. How will you go about achieving your desire results? The answer to this you can call strategy.

- William E Rothschild

✧✧✧

History shows that weakness is provocative.

- Donald Rumsfeld, Secretary of Defence (2001–06)

✧✧✧

I would not say that the future is necessarily less predictable than the past. I think the past was not predictable when it started.

- Donald Rumsfeld

✧✧✧

We do know of certain knowledge that he [Osama Bin Laden] is either in Afghanistan, or in some other country, or dead.

- Donald Rumsfeld

✧✧✧

Reports that say that something hasn't happened are always interesting to me, because as we know, there are known knowns; there are things we know we know. We also know there are known unknowns; that is to say we know there are some things we do not know. But there are also unknown unknowns—the ones we don't know we don't know.

- Donald Rumsfeld
Department of Defence news briefing,
12 February 2002

✧✧✧

There's another way to phrase that and that is that the absence of evidence is not the evidence of absence. It is basically saying the same thing in a different way. Simply because you do not have evidence that something does exist does not mean that you have evidence that it doesn't exist."—on Iraq's weapons of mass destruction.

- Donald Rumsfeld

✧✧✧

If a person is determined to fight to the death, then they may very well have that opportunity.

- Donald Rumsfeld
on Iraqi Resistance Fighters

✧✧✧

If you are not criticized, you may not be doing much.

- Donald Rumsfeld

✧✧✧

Is it likely that an aircraft carrier or a cruise missile is going to find a person?

- US Defence Secretary Donald H Rumsfeld,
regards questions on an air war to kill Osama bin Laden, 23 September 2001.

✧✧✧

To conquer fear is the beginning of wisdom.

- Bertrand Russell

✧✧✧

Do not fear to be eccentric in opinion, for every opinion now accepted was once eccentric.

- Bertrand Russell

✧✧✧

The best tank terrain is that without anti-tank weapons.

- Russian military doctrine

✧✧✧

The greatest contributor to the feeling of tension and fear of war arose from the power of the bombing aeroplane. If all nations would consent to abolish air bombardment . . . that would mean the greatest possible release from fear.

- Ernest Rutherford

✧✧✧

S

If you want to bake an apple pie from scratch, you must first create the Universe.

- Carl Sagan

✧✧✧

I have had dreams and I have had nightmares.
I overcame the nightmares because of the dreams.

- Jonas Salk

✧✧✧

Manoeuvres are often media events rather than full-scale war games.

- Donald W. Sampcor, USMC

✧✧✧

The shovel is the brother to the gun.

- Carl Sandburg, American poet
Spanish-American War veteran

✧✧✧

To delight in war is a merit in the soldier, a dangerous quality in the captain, and a positive crime in the statesman.

- George Santayana

✧✧✧

Those who cannot remember the past are condemned to repeat it.

- George Santayana

✧✧✧

When the rich make war it's the poor that die.

- Jean-Paul Sartre

✧✧✧

Fascism is not defined by the number of its victims, but by the way it kills them.

- Jean-Paul Sartre

✧✧✧

When the rich make war it's the poor that die.

- Jean-Paul Sartre

✧✧✧

'Good-morning; good-morning!' the General said
When we met him last week on our way to the line.
Now the soldiers he smiled at are most of 'em dead,
And we're cursing his staff for incompetent swine.
'He's a cheery old card,' grunted Harry to Jack
As they slogged up to Arras with rifle and pack.
But he did for them both by his plan of attack.

- Siegfried Sassoon (1886–1967) The General

✧✧✧

During the morning we were a silent battalion, except for snoring. Some eight-inch guns were firing about 200 yards from the hollow, but our slumbers were inured to noises which would have kept us wide awake in civilian life. We were lucky to be dry, for the sky was overcast. At one o'clock our old enemy the rain arrived in full force. Four hours' deluge left the troops drenched and disconsolate, and then Dottrell made one of his providential appearances with the rations. Dixies of hot tea, and the rum issue, made all the difference to our outlook.

- Siegfried Sassoon
Memoirs of an Infantry Officer

✧✧✧

We'd gained our first objective hours before
While dawn broke like a face with blinking eyes,
Pallid, unshaved and thirsty, blind with smoke.
Things seemed all right at first. We held their line,
With bombers posted, Lewis guns well placed,
And clink of shovels deepening the shallow trench.
The place was rotten with dead; green clumsy legs
High-booted, sprawled and grovelled along the saps
And trunks, face downward, in the sucking mud,
Wallowed like trodden sand-bags loosely filled;
And naked sodden buttocks, mats of hair,
Bulged, clotted heads slept in the plastering slime.
And then the rain began, the jolly old rain!

A yawning soldier knelt against the bank,
Staring across the morning blear with fog;
He wondered when the Allemands would get busy;
And then, of course, they started with five-nines
Traversing, sure as fate, and never a dud.
Mute in the clamour of shells he watched them burst
Spouting dark earth and wire with gusts from hell,
While posturing giants dissolved in drifts of smoke.
He crouched and flinched, dizzy with galloping fear,

Sick for escape,- loathing the strangled horror
And butchered, frantic gestures of the dead.

An officer came blundering down the trench:
'Stand-to and man the fire-step! 'On he went...
Gasping and bawling, 'Fire- step...counter-attack!'
Then the haze lifted. Bombing on the right
Down the old sap: machine- guns on the left;
And stumbling figures looming out in front.
'O Christ, they're coming at us!' Bullets spat,
And he remembered his rifle...rapid fire...
And started blazing wildly...then a bang
Crumpled and spun him sideways, knocked him
out To grunt and wriggle: none heeded him; he choked
And fought the flapping veils of smothering gloom,
Lost in a blurred confusion of yells and groans...
Down, and down, and down, he sank and drowned,
Bleeding to death. The counter-attack had failed.

- Siegfried Sassoon (1886–1967) - The Counter-Attack

✧✧✧

... to follow the dictum of the British NCO who, when asked where his officers were, replied, 'When it comes time to die, they'll be with us.

- Richard A. Gabriel and Paul L. Savage
Crisis in Command: Mismanagement in the Army

✧✧✧

It is not big armies that win battles, it is the good ones!

- Marshal Maurice de Saxe (Mes Reveries)

✧✧✧

Very few men occupy themselves with the higher problems of war. They pass their lives drilling troops and believe that this is the only branch of the military art. When they arrive at the command of armies they are totally ignorant, and in default of knowing what should be done, they do what they know.

- Marshal Maurice de Saxe (Mes Reveries)

✧✧✧

All the mystery of combat is in the legs; and it is to the legs that we should apply ourselves.

- Marshal Maurice de Saxe

✧✧✧

War is a science covered by darkness, in the obscurity of which one cannot march with a sure step.

- Marshal Maurice de Saxe

✧✧✧

Hope encourages men to endure and attempt everything; in depriving them of it, or in making it too distant, you deprive them of their very soul.

- Marshal Maurice Comte de Saxe,
French Army

✧✧✧

The first of all qualities is courage. Without this the others are of little value, since they cannot be used. The second is intelligence, which must be strong and fertile in expedients. The third is health.

- Marshal Comte de Maurice Saxe

✧✧✧

War is a trade for the ignorant and a service for the expert.

- Marshal Maurice de Saxe

✧✧✧

To affirm that the aeroplane is going to 'revolutionize' naval warfare of the future is to be guilty of the wildest exaggeration.

- Scientific American,
16 July 1910

✧✧✧

Military history, when superficially studied, will furnish arguments in support of any theory.

- Bronsart von Schellendorf,
Imperial German officer and military historian

✧✧✧

It is in wars that we have come to call 'limited wars' that the bargaining appears most vividly and is conducted most consciously. The critical targets in such a war are the mind of the enemy...the threat of violence in reserve is more important than the commitment of force in the field... And, like any bargaining situation, a restrained war involves some degree of collaboration between adversaries.

- Thomas C. Schelling,
Arms and Influence, 1966

✧✧✧

They couldn't hit an elephant at this dist....

- General John Sedgwick, 9 May 1864.
(Killed by a sniper during the U.S. Civil war in the battle of Spotsylvania.)

✧✧✧

If you cross the line we shoot in self-defence, or the mines explode .... If you cross it, then is when the threat is fulfilled, either automatically, if we've rigged it so, or by obligation that immediately becomes due. But we can wait—preferably forever; that is our purpose.

- Thomas Schelling
The Strategy of Conflict, 1960

✧✧✧

Reorientation of Game Theory is explained as follows:

[Y]ou're standing at the edge of a cliff, chained by the ankle to someone else. You'll be released, and one of you will get a large prize, as soon as the other gives in. How do you persuade the other guy to give in, when the only method at your disposal—threatening to push him off the cliff—would doom you both?

Answer: You start dancing, closer and closer to the edge. That way, you don't have to convince him that you would do something totally irrational: plunge him and yourself off the cliff. You just have to convince him that you are prepared to take a higher risk than he is of accidentally falling off the cliff. If you can do that, you win.

- Thomas Schelling

✧✧✧

The question was asked, 'What can seven divisions do to stop a Soviet attack? ... And the answer very explicitly was: They can guarantee that if the Soviets attack and kill or capture 300,000 Americans, the war won't end there.

- Thomas Schelling

✧✧✧

Game theory is usually defined as concerned with how individuals should choose rationally in situations in which the optimal choice for each depends on the choice of another or the choices of others. I have been, I believe, more concerned with how to select a behaviour that will influence another, or others—how to influence others by constraining their expectation of how oneself will act. This has led me to consider the institutional, legal, cultural, sometimes physical circumstances that make it possible to commit and reliably signal promises, threats, negotiating positions, and other kinds of commitments, and to appreciate when inability or disability, even deliberately incurred, may induce the favourable behaviour of others. This is still game theory, but with a different slant.

- Thomas Schelling

✧✧✧

To inflict suffering gains nothing and saves nothing directly; it can only make people behave to avoid it. The only purpose, unless sport or revenge, must be to influence somebody's behaviour, to coerce his decision or choice. To be coercive, violence has to be anticipated. And it has to be avoidable by accommodation. The power to hurt is bargaining power.

- Thomas Schelling
Arms and Influence

✧✧✧

Against stupidity, even the Gods battle in vain.

- Schiller

✧✧✧

The forty-eight hours after the march into the Rhineland were the most nerve-racking in my life. If the French had then marched into the Rhineland, we would have had to withdraw with our tails between our legs, for the military resources at our disposal would have been wholly inadequate for even moderate resistance.

- Dr. Paul Schmidt, Hitler's interpreter

✧✧✧

For anyone who aspires to become a great commander, there is an open book called military history, which began with the hand-to-hand struggle between Cain and Abel and which did not end with the Napoleonic campaigns. .... Its reading, I must admit, is not always exciting. One has to plough through a mass of uninteresting details! But—one accumulates facts , often encouraging facts! And at the bottom of it all, one arrives at the final realization how it all came about, how it had to happen and may happen again.

- Count Alfred von Schlieffen

✧✧✧

To win, we must endeavour to be the stronger of the two at the point of impact. Our only hope of this lies in making our own choice of operations, not in waiting passively for whatever the enemy chooses for us.

- Count Alfred von Schlieffen

✧✧✧

Accomplish much, remain in the background, be more than you appear to be.

- General Field Marshal Count Alfred von Schlieffen

✧✧✧

A complete battle of Cannae is rarely met in history. For its achievement, a Hannibal is needed on the one side, and a Terentius Varro, on the other, both cooperating for the attainment of the great goal.

- General Field Marshal Count Alfred von Schlieffen

✧✧✧

The discipline which makes the soldiers of a free country reliable in battle is not to be gained by harsh or tyrannical treatment. On the contrary, such treatment is far more likely to destroy than to make an army. It is possible to impart instruction and to give commands in such manner and such a tone of voice to inspire in the soldier no feeling but an intense desire to obey, while the opposite manner and tone of voice cannot fail to excite strong resentment and a desire to disobey. The one mode or the other of dealing with subordinates springs from a corresponding spirit in the breast of the commander. He who feels the respect which is due to others cannot fail to inspire in them regard for himself, while he who feels, and hence manifests, disrespect toward others, especially his inferiors, cannot fail to inspire hatred against himself.

- Major General John M. Schofield
Address to the United States Corps of Cadets, 11 August 1879

✧✧✧

I am quite confident that in the foreseeable future armed conflict will not take the form of huge land armies facing each other across extended battle lines, as they did in World War I and World War II or, for that matter, as they would have if NATO had faced the Warsaw Pact on the field of battle.

- General H. Norman Schwarzkopf

✧✧✧

As far as Saddam Hussein being a great military strategist, he is neither a strategist, nor is he schooled in the operational art, nor is he a tactician, nor is he a general, nor is he a soldier. Other than that he's a great military man—I want you to know that.

- General H. Norman Schwarzkopf

✧✧✧

I feel that retired generals should never miss an opportunity to remain silent concerning matters for which they are no longer responsible.

- General H. Norman Schwarzkopf

✧✧✧

If we do go to war, psychological operations are going to be absolutely a critical, critical part of any campaign that we must get involved in.

- General H. Norman Schwarzkopf

✧✧✧

It doesn't take a hero to order men into battle. It takes a hero to be one of those men who goes into battle.

- General H. Norman Schwarzkopf

✧✧✧

Leadership is a potent combination of strategy and character. But if you must be without one, be without the strategy.

- General H. Norman Schwarzkopf

✧✧✧

The truth of the matter is that you always know the right thing to do. The hard part is doing it.

- General H. Norman Schwarzkopf

✧✧✧

In 1848, when I was Brigade-Major at Agra, I made a good many inquiries into the condition of the soldier in barracks, their wants and habits,... One day an intelligent sergeant of the 24th came to me on business, and, amongst other questions, I asked him-"The men, in this hot weather, are confined to barracks from 8 a.m. to 4 p.m. How do they employ themselves during these hours?" "Well, sir, they mainly sit upon their cots and look at each other."

- Major-General Sir Thomas Seaton,
Cadet to Colonel, 1866

✧✧✧

Soldier, rest! thy warfare o'er,
Sleep the sleep that knows not breaking:
Dream of battled fields no more,
Days of danger, nights of waking.
In our isle's enchanted hall,
Hands unseen thy couch are strewing,
Fairy strains of music fall,
Every sense in slumber dewing.
Soldier, rest! thy warfare o'er,
Dream of fighting fields no more:
Sleep the sleep that knows not breaking,
Morn of toil, nor night of waking.

No rude sound shall reach thine ear,
Armour's clang, or war-steed champing,
Trump nor pibroch summon here
Mustering clan, or squadron tramping.
Yet the lark's shrill fife may come
At the day-break from the fallow,
And the bittern sound his drum,
Booming from the sedgy shallow.
Ruder sounds shall none be near,
Guards nor warders challenge here,
Here's no war-steed's neigh and champing,
Shouting clans or squadrons stamping.

Huntsman, rest! thy chase is done,
While our slumbrous spells assail ye,
Dream not, with the rising sun,
Bugles here shall sound reveillé.
Sleep! the deer is in his den;
Sleep! thy hounds are by thee lying;
Sleep! nor dream in yonder glen,
How thy gallant steed lay dying.
Huntsman, rest; thy chase is done,
Think not of the rising sun,
For at dawning to assail ye,
Here no bugles sound reveillé.

- Sir Walter-Scott
The Lady of the Lake; Soldier, Rest! Thy Warfare O'er

✧✧✧

It is not because things are difficult that we do not dare.
It is because we do not dare that things are difficult.

- Seneca

✧✧✧

The fear of war is worse than war itself.

- Seneca

✧✧✧

Courage is sustained...by calling up anew the vision of the goal.

- A. G. Sertillanges

✧✧✧

Air power speaks a strategic language so new that translation into the hackneyed idiom of the past is impossible.

- Alexander de Seversky (Victory through Air Power)

✧✧✧

Only air power can defeat air power. The actual elimination or even stalemating of an attacking air force can be achieved only by a superior air force.

- Major Alexander P. de Seversky, USAAF

✧✧✧

... under the German version of directive control, the subordinate is free to modify the task set him without referring back, if he is satisfied that further pursuit of that aim would not represent the best use of his resources in furtherance of his superior's intention."

- Richard E. Simpkin; Race to the Swift - Thoughts on Twenty-First Century Warfare, 1985

✧✧✧

As long as he retains the initiative, the attacker can get away with this by exploiting momentum—a kind of steam-roller effect. The defender by contrast must combine speed with precision in his response. Conditioned as it may be by extraneous factors, the way the Soviet Army seems to have gone in exploiting technology is just one example of the misuse of computer technology.

- Richard E. Simpkin; Race to the Swift - Thoughts on Twenty-First Century Warfare, 1985

✧✧✧

The answer lies in Standard Operating Procedures. But it has to be a carefully balanced answer. The scope and nature of Standard Operating Procedures are critical, the attitudes of commanders and staff towards them les so—and the same goes for C^3 technology! In a word, Standard Operating Procedures must be slaves restricted to servile tasks. They must not have free men's tasks entrusted to them. Above all, they must never become masters.

- Richard Simpkin (Race to the Swift)

✧✧✧

Faced with the same new situation, the immediate reaction of the British soldier, whether army or section commander, was to refer to higher headquarters, seek additional information and try to plan a reasoned response. A German soldier tended to respond to the same situation by rapidly deciding what action would best fit into the overall concept of the operation and immediately implementing it.

- Richard Simpkin (Race to the Swift)

✧✧✧

At the same time, within the limits of the task assigned him, every commander must show all possible independence, boldness, and resourcefulness. And again, within the limits of his task, each commander is responsible for the control of his subordinates to the point where he must concentrate all his initiative and resourcefulness on this. On the other hand, every junior commander must have a good understanding not just of his own task but of his superiors' general mission and must keep himself in the broader picture. In this way, in the event of an unexpected breakdown of all communications with his superior commander, he will be in a position to take independent decisions which conform as closely as possible to the general situation.

- Richard Simpkin,
quoted from Questions of Higher Command (from the book of that name published in 1924), Deep Battle:
The Brainchild of Marshal Tukhachevskii, 1987

✧✧✧

The combination of rain and movement produces mud, which has a general degrading effect on men and machines and may seriously damage some types of equipment. More important from the operational viewpoint, rain can rather quickly render impassable areas of marginal going in general, and the approaches to and exits from water-obstacle crossings in particular.

- Richard E. Simpkin
Race to the Swift - Thoughts on Twenty-First Century Warfare

✧✧✧

No amount of modernisation of arms, equipment, tactics and organisations can produce results unless we have the right kind of man in the right state of mind, manning the system.

- General K Sundarji, Chief of Army Staff,
India (1986-1988)

✧✧✧

The bed-rock of elan is the professional competence of individuals and leaders, and the faith, confidence and pride in the effectiveness of the group—the section upwards, to the Army as a whole.

- General K Sundarji, Chief of Army Staff,
India (1986-1988)

✧✧✧

In peace nothing so becomes a man as modest stillness and humility; but when the blast of war blows in our ears, then imitate the action of the tiger; stiffen the sinews, disguise fair nature with hard favour'd rage. . .

- William Shakespeare (Henry V)

✧✧✧

Our legions are brimful, our cause is ripe:
The enemy increaseth every day;
We, at the height, are ready to decline.
There is a tide in the affairs of men which taken at the flood, leads on to fortune;
Omitted, all the voyage of their life is bound in shallows and in miseries;
On such a full sea are we now afloat;
And we must take the current when it serves or lose our ventures.

- William Shakespeare (Julius Caesar)

✧✧✧

Cowards die many times before their deaths; the valiant never taste death but once.

- William Shakespeare ("Julius Caesar")

✧✧✧

In peace nothing so becomes a man as modest stillness and humility; but when the blast of war blows in our ears, then imitate the action of the tiger; stiffen the sinews, disguise fair nature with hard favoured rage ...

We few, we happy few, we band of brothers. For he today that sheds his blood with me, Shall be my brother; be ne'er so vile,

This day shall gentle his condition. And gentlemen in England now abed, Shall think themselves accursed they were not here,

And hold their manhood's cheap whiles any speaks, That fought with us upon Saint Crispin's day ... From now until the end of the world, we and it shall be remembered. We few, we band of brothers. For he who sheds his blood with me shall be my brother.

- William Shakespeare (Henry V)
reminding his men of the importance of their comradeship before the battle of Agincourt.

✧✧✧

There is a tide in the affairs of men.
Which, taken at the flood, leads on to fortune;
Omitted, all the voyage of their life
Is bound in shallows and in miseries.
On such a full sea are we now afloat,
And we must take the current when it serves,
Or lose our ventures.

- William Shakespeare (Julius Caesar)

✧✧✧

Cry 'Havoc' and let slip the dogs of War.

- William Shakespeare (Julius Caesar)

✧✧✧

The enemies are only 50 yards from us. We are heavily outnumbered. We are under devastating fire. I shall not withdraw an inch but will fight to our last man and our last round.

- Major Somnath Sharma, the first winner of the Param Vir Chakra, awarded posthumously (last communication on radio)

✧✧✧

You can always tell an old soldier by the inside of his holsters and cartridge boxes. The young men carry pistols and cartridges; the old ones, grub.

- George Bernard Shaw

✧✧✧

We learn from history that we learn nothing from history.

- George Bernard Shaw

✧✧✧

The capacity of any conqueror is more likely than not to be an illusion produced by the incapacity of his adversary.

- George Bernard Shaw

✧✧✧

When a stupid man is doing something he is ashamed of, he always declares that it is his duty.

- George Bernard Shaw

✧✧✧

When the military man approaches, the world locks up its spoons and packs off its womankind.

- George Bernard Shaw

✧✧✧

There's only one truth about war: people die.

- Sheridan

✧✧✧

War is cruelty. There's no use trying to reform it, the crueller it is the sooner it will be over.

- William Tecumseh Sherman

✧✧✧

There is many a boy here today who looks on war as all glory, but, boys, it is all hell.

- General William T. Sherman, 1880

✧✧✧

Every day I feel more and more in need of an atlas, as the knowledge of geography in its minute details is essential to a true military education.

- Lieutenant General William Tecumseh Sherman

✧✧✧

Grasping the special qualities of terrain can decide the battle for a general and understanding that same terrain is the key for a historian's later understanding of that same battle.

- Lieutenant General William Tecumseh Sherman

✧✧✧

War is the remedy that our enemies have chosen, and I say let us give them all they want.

- General William T. Sherman

✧✧✧

Fighting and winning our nation's wars, this is our nonnegotiable contract.

- General Eric K. Shinseki, United States Army Chief of Staff

✧✧✧

If you believe in the true Divine Book and the Word of God (the Quran), you will find there *Rabb-ul-Alamin*, the Lord of all men, and not *Rabb-ul-Muslimin*, the Lord of the Muslims only. Islam and Hinduism are terms in contrast. They are (diverse pigments) used by the true Divine Painter for blending the colours and filling in the outlines .... If it be a mosque, the call to prayer is chanted in remembrance of Him. If it be a temple, the bell is rung in yearning for Him only. To show bigotry for any man's own creed and practices is equivalent to altering the words of the Holy Book.

- Shivaji, Ruler of the Maratha Kingdom
In a letter to the Mughal Emperor, Aurangzeb

✧✧✧

He'll never make a good soldier should not be heard from the confident leader. Seemingly incompetent men should be a challenge to one's ability. They should be studied more closely and the best method of handling them discovered. In every man there is much that may be made useful for military purposes.

- "Discipline and Personality," by Sergeant Major E.J. Simon, R.C.R., Canadian Defence Quarterly, Vol. II, No. 3, April, 1925

✧✧✧

Chun ka azhameh heelate dar guzasht;
Har haal tey darguzhast;
Halal ast burdan;
Ba shamsheer dast

When all avenues have been tried;
Yet justice is not in sight;
It is right to pick up the sword;
It is then right to fight.

- Guru Gobind Singh
from the Zafarnama (Epistle of Victory)

✧✧✧

Deh Shiva var mohe ehae,
shubh karman se kabhoon na taroon;
Na daroon ar son jab jaiye laroon,
nishchay kar apni jeet karoon;
Ar sikh hoon apne hi man ko,
el lalach ho gun toh uchron;
Jab auv ki audh nidhan bane,
at hi ran mein tab jujh maroon.

O Lord grant me the boon,
that I may never deviate from doing a good deed.
That I shall not fear when I go into combat.
And with determination I will be victorious.
That I may teach myself this greed alone,
to learn only Thy praises.
And when the last days of my life come,
I may die in the midst of battle.

- Guru Gobind Singh
Chandi Charitar Ukti Bilas, in the Dasam Granth

✧✧✧

Identifying your enthusiasms requires courage and heroic creative vision. You have to believe that what you want is possible for you.

- Marsha Sinetar

✧✧✧

A good inter-Service staff officer must first be a good officer of his own Service, and we should lose more than we gained by merging the identity of the three Staff Colleges.

- Marshal of the Royal Air Force John Slessor
(The Central Blue)

✧✧✧

If there is one attitude more dangerous than to assume that a future war will be just like the last one, it is to imagine that it will be so utterly different we can afford to ignore all the lessons of the last one.

- John C. Slessor (Air Power and Armies)

✧✧✧

It is better to be wise after the event than not wise at all, and wisdom after one event may lead to wisdom before another.

- John C. Slessor (Air Power and Armies)

✧✧✧

The real importance of Munich . . . is that it was the first victory for air power-no less significant for being temporarily bloodless.

- Marshal of the Royal Air Force John Slessor

✧✧✧

An air force commander must exploit the extreme flexibility, the high tactical mobility, and the supreme offensive quality inherent in air forces, to mystify and mislead his enemy, and so to threaten his various vital centres as to compel him to be dangerously weak at the point which is *really decisive* at the time.

- Marshal of the Royal Air Force John Slessor

✧✧✧

All men have some degree of physical courage—it is surprising how much. Courage, you know, is like having money in the bank; we start with a certain capital of courage, some large, some small, and we proceed to draw on our balance, for, don't forget, courage is an expendable quality. We can use it up. If there are heavy and, what is more serious, if there are continuous calls on our courage, we begin to overdraw. If we go on overdrawing, we go bankrupt—we break down.

- General Sir William Slim, Courage and Other Broadcasts

✧✧✧

The soldier's actions are scrutinized by critics rather unfairly. He has to make a vital decision on the battlefield on incomplete information in a matter of seconds and afterwards, the experts can sit down at leisure, with all the facts before them and argue about what he might, could or should have done, and he is lucky if tactical experts decide after twenty years of profound consideration that what he did in three minutes was right and the critics will twist, falsify, misinterpret or invent any fact of evidence to prove he is an inhuman monster wallowing in innocent blood.

- General William Slim

✧✧✧

Preparation for war is an expensive, burdensome business, yet there is one important part of it that costs little—study.

- General William Slim

✧✧✧

If you want to talk to men, it doesn't matter whether they are private soldiers or staff officers, if you want to talk to them as a soldier, and not as a politician, there are only two things necessary. The first is to have something to say that is worth saying, to know what you want to say: and the second, and terribly important thing, is to believe in yourself. Don't go and tell men something you don't believe yourself, because they'll spot it and if they don't spot it at the time, they'll find out. Then you're finished.

- Gen. Sir William Slim (Viscount Slim)

✧✧✧

Everything that is shot or thrown at you or dropped on you in war is most unpleasant but of all horrible devices, the most terrifying. . . is the land mine.

- Sir William Slim,
Unofficial History, 1959

✧✧✧

In battle nothing is ever as good or as bad as the first reports of excited men would have it.

- Field Marshal William Slim

✧✧✧

Officers are there to lead. As officers, you can neither eat, nor drink, nor sleep, nor smoke, nor even sit down until you have personally seen that your men have done these things. If you will do this for them, they will follow you to the end of the world.

- Field Marshal William Slim

✧✧✧

When you cannot make up your mind which of two evenly balanced courses of action you should take—choose the bolder.

- Field Marshal William Slim

✧✧✧

Personal leadership exists only as the officers demonstrate it by superior courage, wider knowledge, quicker initiative, and a greater readiness to accept responsibility than those they lead.

- Field Marshal Sir William Slim

✧✧✧

The foundations of morale are, I think, first, spiritual, then mental, and lastly material. I put them in that order because that, I believe, is the order of their importance.

- Field Marshal Sir William Slim,
Chief of the Imperial General Staff

✧✧✧

There is no beating these troops in spite of their generals. I always thought them bad soldiers, now I am sure of it. I turned their right, pierced their centre, broke them everywhere; the day was mine, and yet they did not know it and would not run.

- Marshal Soult, Albuhera 1811

✧✧✧

Neither bars nor stars make an officer. An individual becomes an officer only when he develops those inner qualities of honesty, self-sacrifice, and attention to duty that are always inherent to real leadership.

- General Samuel D. Sturgis, Jr

✧✧✧

Nothing is more useful than water; but it will purchase scarce nothing; scarce anything can be had in exchange for it. A diamond, on the contrary, has scarce any value in use; but a very great quantity of other goods may frequently be had in exchange for it.

- Adam Smith (Wealth of Nations, 1776)

✧✧✧

Most of the time, leaders should laugh at themselves rather than others.

- Major General Perry M. Smith, U.S. Air Force

✧✧✧

Air battle is not decided in a few great clashes but over a long period of time when attrition and discouragement eventually cause one side to avoid the invading air force.

- Dale O. Smith

✧✧✧

There will be demands upon your ability, upon your endurance, upon your disposition, upon your patience...just as fire tempers iron into fine steel so does adversity temper one's character into firmness, tolerance and determination.

- Senator Margaret Chase Smith
Lt Colonel, United States Air Force Reserve

✧✧✧

The test of courage comes when we are in the minority. The test of tolerance comes when we are in the majority.

- Ralph W. Sockman

✧✧✧

The general must know how to get his men their rations and every other kind of stores needed for war. He must have imagination to originate plans- practical sense and energy to carry them through. He must be: observant, untiring, shrewd, kindly and cruel, simple, and crafty, a watchman and a robber, lavish and miserly, generous and stingy, rash and conservative. All these and many other qualities both natural and acquired he must have. He should also, as a matter of course, know his tactics; for a disorderly mob is no more an army than a heap of building materials is a house.

- Socrates

✧✧✧

Courage is endurance of the soul.

- Socrates

✧✧✧

In our victory over Japan, airpower was unquestionably decisive. That the planned invasion of the Japanese Home islands was unnecessary is clear evidence that airpower has evolved into a force in war co-equal with land and sea power, decisive in its own right and worthy of the faith of its prophets.

- General Carl A. 'Tooey' Spaatz, 'Evolution of Air Power,'
Military Review, 1947.

✧✧✧

The first and absolute requirement of strategic air power in this war was control of the air in order to carry out sustained operations without prohibitive losses.

- General Carl A. 'Tooey' Spaatz

✧✧✧

Air control can be established by superiority in numbers, by better employment, by better equipment, or by a combination of these factors.

- General Carl A. 'Tooey' Spaatz

✧✧✧

As things heated up in the Pacific, Japanese leaders asked, "Where are the battleships?" We better be prepared to dominate the skies above the surface of the earth or be prepared to be buried beneath it.

- General Carl A. "Tooey" Spaatz

✧✧✧

Air control can be established by superiority in numbers, by better employment, by better equipment, or by a combination of these factors.

- General Carl A. 'Tooey' Spaatz

✧✧✧

It takes close cooperation with the Army to obtain the maximum misuse of air power.

- Gen Carl Spaatz

✧✧✧

The strain on our military personnel and their families continues to grow as the services are being asked to do more with less, while the perennial promise of adequate budgets continues to be pushed further out into the future.

- Rep. Floyd D. Spence
Chairman of the House National Security Committee, 4 March 1996

✧✧✧

It isn't just my brother's country, or my husband's country, it's my country as well. And so the war wasn't just their war, it was my war and I needed to serve in it.

- Beatrice Hood Stroup, Major, Woman's Army Corps, WWII

✧✧✧

The bitterest tears shed over graves are for words left unsaid and deeds left undone.

- Harriet Beecher Stowe

✧✧✧

Peace is not an absence of war, it is a virtue, a state of mind, a disposition for benevolence, confidence, justice.

- Baruch Spinoza

✧✧✧

A sincere diplomat is like dry water or wooden iron.

- Stalin, 1913

✧✧✧

The death of one man is tragic, but the death of thousands is statistic.

- Joseph Stalin

✧✧✧

The higher a monkey climbs, the more you see its behind.

- General Joseph Stilwell

✧✧✧

Good eating, if you are hungry.

- General Joseph Stilwell (After a thoughtful pause, while delivering a lecture on the Sino-Japanese war, in response to a cavalry officer who asked a question on the role of the horse in the fighting in China)

✧✧✧

To,
The Commander Jamalpur Garrison

I am directed to inform you that your garrison has been cut off from all sides and you have no escape route available to you. One brigade with full complement of artillery has already been built up and another will be striking by morning. In addition you have been given a foretaste of a small element of our air force with a lot more to come. The situation as far as you are concerned is hopeless. Your higher commanders have already ditched you.

I expect your reply before 6.30 p.m. today failing which I will be constrained to deliver the final blow for which purpose 40 sorties of MiGs have been allotted to me.

In this morning's action the prisoners captured by us have given your strength and dispositions, and are well looked after.

The treatment I expect to be given to the civil messenger should be according to a gentlemanly code of honour and no harm should come to him

An immediate reply is solicited.

Brigadier HS Kler, Commander

(*The reply was sent a few hours later via the same messenger (the reply letter also included a bullet*)

Dear Brigadier,

Hope this finds you in high spirits. Your letter asking us to surrender had been received. I want to tell you that the fighting you have seen so far is very little, in fact the fighting has not even started. So let us stop negotiating and start the fight.

40 sorties, I may point out, are inadequate. Ask for many more. Your point about treating your messenger well was superfluous. It shows how you under-estimate my boys. I hope he liked his tea.

Give my love to the Muktis. Let me see you with a sten in your hand next time instead of the pen you seem to have such mastery over.

Now get on and fight.

Yours sincerely

- (Lt. Colonel Ahmed Sultan)
Commander Jamalpur fortress

✧✧✧

The limey layout is simply stupendous, you trip over Lieutenant-Generals on every floor, most of them doing captains work, or none at all.

- General Joseph Stillwell, about the Burma campaign 1943

✧✧✧

More than one general has redeemed faulty dispositions and won fame by a suitably glorious death.

- James Lawton Stokesbury, military historian

✧✧✧

In the history of the world, no one has ever washed a rented car.

- Larry Summers

✧✧✧

We have enough religion to make us hate, but not enough to make us love one another.

- Jonathan Swift

✧✧✧

War: that mad game the world so loves to play.

- Jonathan Swift

✧✧✧

One enemy can do more hurt than ten friends can do good.

- Jonathan Swift, 1667–1745

✧✧✧

Service aviation must be the spearhead, civil aviation the shaft, of our air effort.

- Maj Gen F.H. Sykes, RAF, 1922

✧✧✧

T

The modern industrial society cannot function, and its government cannot govern, except with popular participation and by popular consent.... They must make great concessions to popular notions of what is democratic and just, or be replaced by regimes that will do so.... They must use the liberal rhetoric and also pay something in the way of social compromise...if they are to retain power and keep the people to their accustomed, profit-producing tasks. This fact makes such governments extremely vulnerable to a sort of war—guerrilla war with its psychological and economic weapons—that their predecessors could have ignored, had such a war been possible at all in the past. They are vulnerable because they must, at all cost, keep the economy functioning and showing a profit.... Again, they are vulnerable because they must maintain the appearance of normalcy; they can be embarrassed out of office. And they are triply vulnerable because they cannot be as ruthless as the situation demands. They cannot openly crush the opposition that embarrasses and harasses them. They must be wooers as well as doers.

- Robert Taber, The War of the Flea

✧✧✧

Whether the primary cause of revolution is nationalism, or social justice, or the anticipation of material progress, the decision to fight and to sacrifice is a social and a moral decision. Insurgency is thus a matter not of manipulation but of inspiration.

I am aware that such conclusions are not compatible with the pictures of guerrilla operations and guerrilla motivations drawn by the counterinsurgency theorists who are so much in vogue today. But the counterinsurgency experts have yet to win a war. At this writing, they are certainly losing one.

- Robert Taber, War of the Flea

✧✧✧

Their picture is distorted because their premises are false and their observation faulty. They assume–perhaps their commitments require them to assume–that politics is mainly a manipulative science and insurgency mainly a politico-military technique to be countered by some other technique; whereas both are forms of social behaviour, the latter being the mode of popular resistance to unpopular governments.

- Robert Taber, War of the Flea

✧✧✧

Conditions have changed in the world. What is wanted today is manpower and its products. The raw materials of the undeveloped areas are of no use to the industrial powers... without the human effort that makes them available; strategic bases require the services and the good will of large populations; industry requires both large labour pools and ever-expanding consumer markets.

- Robert Taber, War of the Flea

✧✧✧

Under such conditions, to try to suppress popular resistance movements by force is futile. If inadequate force is applied, the resistance grows. If the overwhelming force necessary to accomplish the task is applied, its object is destroyed. It is a case of shooting the horse because he refuses to pull the cart.

- Robert Taber, War of the Flea

✧✧✧

Analogically, the guerrilla fights the war of the flea, and his military enemy suffers the dog's disadvantages: too much to defend; too small, ubiquitous, and agile an enemy to come to grips with. If the war continues long enough—this is the theory—the dog succumbs to exhaustion and anaemia without ever having found anything on which to close its jaws or to rake with its claws.

- Robert Taber, War of the Flea

✧✧✧

Reason and calm judgement—the qualities specially belong to a leader.

- Tacitus, 55-177

✧✧✧

There are only two kinds of people on this beach; the dead and those about to die. So let's get the hell out'a here.

- Colonel George Taylor
Commander of the 16th Infantry, Omaha beach, 6 June 1944

✧✧✧

For the essence of war is not to take this point or that, but to destroy, or at least weaken, the military strength of the enemy.

- A.J.P. Taylor, Rumours of Wars

✧✧✧

Nothing is inevitable until it happens.

- A. J. P. Taylor

✧✧✧

Human blunders usually do more to shape history than human wickedness.

- A. J. P. Taylor

✧✧✧

No matter what political reasons are given for war, the underlying reason is always economic.

- A. J. P. Taylor

✧✧✧

Half a league, half a league,
Half a league onward,
All in the valley of Death
Rode the six hundred.
"Forward the Light Brigade!
Charge for the guns!" he said.
Into the valley of Death
Rode the six hundred.

Forward, the Light Brigade!"
Was there a man dismay'd?
Not tho' the soldier knew
Someone had blunder'd.
Theirs not to make reply,
Theirs not to reason why,
Theirs but to do and die.
Into the valley of Death
Rode the six hundred.

Cannon to right of them,
Cannon to left of them,
Cannon in front of them
Volley'd and thunder'd;
Storm'd at with shot and shell,
Boldly they rode and well,
Into the jaws of Death,
Into the mouth of hell
Rode the six hundred.

Flash'd all their sabres bare,
Flash'd as they turn'd in air
Sabring the gunners there,
Charging an army, while
All the world wonder'd.
Plunged in the battery-smoke
Right thro' the line they broke;
Cossack and Russian
Reel'd from the sabre-stroke
Shatter'd and sunder'd.
Then they rode back, but not,
Not the six hundred.

Cannon to right of them,
Cannon to left of them,
Cannon behind them
Volley'd and thunder'd;
Storm'd at with shot and shell,
While horse and hero fell,
They that had fought so well
Came thro' the jaws of Death,
Back from the mouth of hell,
All that was left of them,
Left of six hundred.

When can their glory fade?
O the wild charge they made!
All the world wonder'd.
Honour the charge they made!
Honour the Light Brigade,
Noble six hundred!

- Alfred Tennyson (The Charge of the Light Brigade)

✧✧✧

He makes no friend who never made a foe.

- Alfred Tennyson

✧✧✧

If we have no peace, it is because we have forgotten that we belong to each other.

- Mother Teresa

✧✧✧

Quintus Septimius Florens Tertullianus

He who flees will fight again .....

- Tertullian (160 A.D.)

✧✧✧

You may have to fight a battle more than once to win it.

- Margaret Thatcher

✧✧✧

There is a difference between recognizing reality and legitimizing it.

- Margaret Thatcher

✧✧✧

Guerrillas never win wars, but their adversaries often lose them.

- Charles W. Thayer

✧✧✧

He who has command of the sea has command of everything.

- Themistocles
Athenian General 524–459 B.C.

✧✧✧

I choose the likely man in preference to the rich man; I want a man without money rather than money without a man.

- Themistocles

✧✧✧

I have with me two gods, Persuasion and Compulsion.

- Themistocles

✧✧✧

Government is at best but an expedient; but most governments are usually, and all governments are sometimes, inexpedient. The objections which have been brought against a standing army, and they are many and weighty, and deserve to prevail, may also at last be brought against a standing government.

- Henry David Thoreau

✧✧✧

Thank God, men cannot as yet fly, and lay waste the sky as well as the earth.

- Henry David Thoreau, 3 January 1861

✧✧✧

He who learns and runs away, lives to learn another day.

- Edward Lee Thorndike

✧✧✧

The entropy of any closed system always tends to increase, and thus the nature of any given system is continuously changing even as efforts are directed toward maintaining it in its original form.

- Second Law of Thermodynamics

✧✧✧

The Nation that makes a great distinction between its scholars and its warriors will have its thinking done by cowards and its fighting done by fools.

- Thucydides

✧✧✧

Politics among City-states is an effectuation of their own interests based upon consideration of their respective powers. In these considerations, each states' gains are exactly balanced by the losses of others .... Justice and politics comprise two different worlds where the operational rules are fundamentally different. Justice has any meaning among states if power is equally distributed.

- Thucydides

✧✧✧

You, by giving in, would save yourselves from disaster ... [for] your actual resources are too scanty to give you a chance of survival against the forces that are opposed to you at this moment. You will therefore be showing an extraordinary lack of common sense if ... you still fail to reach a conclusion wiser than anything you have mentioned so far ... Think it over again ... and let this be a point that constantly recurs to your minds—that you are discussing the fate of your country, that you have only one country, and its future for good or ill depends upon this one single decision which you are going to make.

- Thucydides, History of Peloponnesian War
Letter from the powerful Athenians to intimidate the Melians

✧✧✧

We used to think that our neutrality was a wise thing, since it prevented us being dragged into danger by other people's policy; now we see it clearly as a lack of foresight and as a source of weakness ... We recognize that we have nothing but our own resources, it is impossible for us to survive, and we can imagine what lies in store for us if they overpower us. We are, therefore, forced to ask for assistance, both from you and from everyone else; and it should not be held against us that now we have faced the facts and are reversing our old policy of keeping ourselves to ourselves.

- Thucydides, History of Peloponnesian War
Representatives from Corcyra approaching Athens

✧✧✧

. . . it was by courage, sense of duty, and a keen feeling of honour in action that men were enabled to win all this, and that no personal failure in an enterprise could make them consent to deprive their country of their valour, but they laid it at her feet as the most glorious contribution that they could offer.

- Thucydides
(The Funeral Oration of Pericles) The History of the Peloponnesian War

✧✧✧

Human history is a cycle which excess of power keeps revolving.

- Thucydides

✧✧✧

Be convinced that to be happy means to be free and that to be free means to be brave. Therefore do not take lightly the perils of war.

- Thucydides

✧✧✧

I shall be satisfied if my words are judged useful by those who desire a clear understanding of the events which occurred in the past and which will occur again, in much the same way in future, human nature being what it is.

- Thucydides

✧✧✧

When a nation loses it's military spirit, the career of arms immediately ceases to be respected and military men drop down to the lowest rank among public officials. They are neither greatly esteemed nor greatly understood .... it is not the leading citizens, but the least important who go into the army ..... The elite of the nation avoid a military career because it is not held in honour, and it is not held in honour because the elite of the nation do not take it up.

- Alexis De Tocqueville

✧✧✧

There is, therefore, no reason for surprise if democratic armies are found to be restless ... The soldier feels he is in a position of inferiority and his wounded pride will give him a taste of war which will make him needed, in the course of which he hopes to win by force of arms, the political influence and personal consideration which have not come his way.

- Alexis De Tocqueville

✧✧✧

Democracy throws mediocrity into power.

- Alexis De Tocqueville

✧✧✧

If your sword is too short, take one step forward.

- Admiral Marquis Heihachiro Togo (1846–1934)

✧✧✧

It is not possible . . . to concentrate enough military planes with military loads over a modern city to destroy that city.

- US Colonel John W. Thomason Jr., November 1937.

✧✧✧

You need three things to win a war, Money, Money and more Money

- Travulzio 15th Century Italian General

✧✧✧

Without war no State could be. All those we know of arose through war, and the protection of their members by armed force remains their primary and essential task. War, therefore, will endure to the end of history, as long as there is a multiplicity of states.

- Heinrich von Treitschke

✧✧✧

You may not be interested in war but war is interested in you.

- Leon Trotsky

❖❖❖

The atom bomb was no 'great decision.' . . . It was merely another powerful weapon in the arsenal of righteousness.

- President Harry Truman

❖❖❖

If you cannot convince them, confuse them.

- President Harry Truman

❖❖❖

A leader is a man who had the ability to get other people to do what they don't want to do, and like it.

- President Harry Truman

❖❖❖

Carry the battle to them. Don't let them bring it to you. Put them on the defensive and don't ever apologize for anything.

- President Harry Truman

❖❖❖

Men make history and not the other way around. In periods where there is no leadership, society stands still. Progress occurs when courageous, skilful leaders seize the opportunity to change things for the better.

- President Harry Truman

❖❖❖

The buck stops here!

- President Harry Truman

❖❖❖

If you can't convince them, confuse them.

- President Harry Truman

❖❖❖

Men don't change. The only thing new in the world is the history you don't know.

- President Harry Truman

❖❖❖

Whether in an advantageous position or a disadvantageous one, the opposite state should be always present to your mind.

- Ts'ao Kung

❖❖❖

Bravery without forethought causes a man to fight blindly and desperately like a mad bull. Such an opponent must not be encountered with brute force, but may be lured into an ambush and slain.

- Ts'ao Kung

❖❖❖

The speed at which a heavy attack can be developed is in itself a surprise; and surprise is in itself a protection, a form of security lasting as long as that surprise and its successors are maintained.

- Maj Gen Francis Tuker

✧✧✧

Small units cannot afford to 'wait for orders'; nor do they have the right to do so. They must act boldly and decisively on their own initiative. It is in reliance on this spirit of initiative and acting without orders that the commander planning the battle issues his orders and directs the action.

- Marshal Tukhachevskii

✧✧✧

If we go on explaining we shall cease to understand one another.

- Charles Maurice de Talleyrand

✧✧✧

Mistrust first impulses; they are nearly always good.

- Charles Maurice de Talleyrand

✧✧✧

Ones reputation is like a shadow, it is gigantic when it precedes you, and a pigmy in proportion when it follows.

- Charles Maurice de Talleyrand

✧✧✧

Speech was given to man to disguise his thoughts.

- Charles Maurice de Talleyrand

✧✧✧

The art of statesmanship is to foresee the inevitable and to expedite its occurrence.

- Charles Maurice de Talleyrand

✧✧✧

They have learned nothing and forgotten nothing.

- Charles Maurice de Talleyrand

✧✧✧

To succeed in the world, it is much more necessary to possess the penetration to discern who is a fool, than to discover who is a clever man.

- Charles Maurice de Talleyrand

✧✧✧

Don't go the way the world takes you, take the world the way you go.

- Mathew Thomas

✧✧✧

My experience taught me that a country's armed services should be an instrument, but never an arbiter, of national policy. The history of countries whose military men were not taught to be apolitical shows what disruptive influences they sometimes can be.

- General KS Thimayya
Quoted in 'Thimayya of India' by Humphrey Evans

✧✧✧

A soldier's problems today stem from the fact that, now, the irresponsible use of military force could destroy the human race. A soldier therefore has greater responsibility to society than he ever had before in history and his duty is to learn to carry that well. To this end, his experience, or training, must make him into the kind of citizen who is beyond narrow attachments to class and province and above the passions of political conflict.

- General KS Thimayya
Quoted in 'Thimayya of India' by Humphrey Evans

✧✧✧

Luck! It's the most important quality a soldier can have.

- General KS Thimayya

✧✧✧

Every act of courage is a manifestation of the ground of being, however questionable the content of the act may be.

- Paul Tillich

✧✧✧

The future has a notorious habit of arriving too soon, and in the wrong order.

- Alvin Toffler

✧✧✧

The illiterate of the 21st century will not be those who cannot read or write, but those who cannot learn, unlearn and re-learn.

- Alvin Toffler

✧✧✧

If the tools of warfare are no longer tanks and artillery, but rather computer viruses and microrobots, then we can no longer say that nations are the only armed groups or that soldiers are the only ones in possession of the tools of war.

- Alvin & Heidi Toffler

✧✧✧

The most important form of property is now intangible. It is super-symbolic. It is knowledge. It reduces the need for raw materials, labour, time, space, capital and other inputs.

- Alvin & Heidi Toffler

✧✧✧

Not all who wander are lost.

- J.R.R. Tolkien

✧✧✧

Without war no State could be. All those we know of arose through war, and the protection of their members by armed force remains their primary and essential task. War, therefore, will endure to the end of history, as long as there is a multiplicity of states.

- Heinrich von Treitschke

✧✧✧

The development of air power in its broadest sense, and including the development of all means of combating missiles that travel through the air, whether fired or dropped, is the first essential to our survival in war.

- Viscount Hugh M. Trenchard, 1946

✧✧✧

I do not rule Russia. 10,000 clerks do.

- Tsar Nicholas I

✧✧✧

Dead battles, like dead generals, hold the military mind in their dead grip and Germans, no less than other peoples, prepare for the last war.

- Barbara W. Tuchman

✧✧✧

War is the unfolding of miscalculations.

- Barbara Tuchman: The Guns of August

✧✧✧

Nothing so comforts the military mind as the maxim of a great but dead general.

- Barbara Tuchman (1912–1989)

✧✧✧

The unrecorded past is none other than our old friend, the tree in the primeval forest which fell without being heard.

- Barbara W. Tuchman

✧✧✧

Honour wears different coats to different eyes.

- Barbara W. Tuchman

✧✧✧

The fleet sailed to its war base in the North Sea, headed not so much for some rendezvous with glory as for rendezvous with discretion.

- Barbara W. Tuchman

✧✧✧

You must love soldiers in order to understand them, and understand them in order to lead them.

- Henri Turenne

✧✧✧

You can't describe the moral lift, when in the fight your spirits weary hears above the hostile fire, your own artillery.
Shells score the air like wavy hair from a forward battery.
As regimental cannon crack while from positions further back,
in bitter sweet song overhead crashing discordantly
Division's pounding joins the attack;
Mother like she belches shell; Glorious it flies, and well,
As, with a hissing screaming squall,
A roaring furnace, giving all, she sears a path for the infantry...."

- Aleksander Tvardovskiy, from the poem "Vasily Tyorkin" 1943

✧✧✧

The rule is perfect: in all matters of opinion, our adversaries are insane

- Mark Twain

✧✧✧

Courage is resistance to fear, mastery of fear—not absence of fear.

- Mark Twain

✧✧✧

We should be careful to get out of an experience only the wisdom that is in it—and stop there; lest we be like the cat that sits down on a hot stove lid. She will never sit down on a hot stove lid again—and that is well; but also she will never sit down on a cold one.

- Mark Twain
Following the Equator: A Journey Around the World

✧✧✧

If our air forces are never used, they have achieved their finest goal.

- General Nathan F. Twining

✧✧✧

If a man who serves indolently and a man who serves well are treated in the same way, the man who serves well may begin to wonder why he does so.

- Asakura Toshikage (1428–81)

✧✧✧

Without harmony in the State, no military expedition can be undertaken; without harmony in the army, no battle array can be formed.

- Wu Tzu

✧✧✧

Weapons are baleful instruments, strife is antagonistic to virtue, a military commander is the negation of civil order!

- Wei Liao Tzu

✧✧✧

He who knows others is wise. He who knows himself is enlightened. He who conquers others has physical strength. He who conquers himself is strong.

- Lao Tzu (Chinese Philosopher, founder of Taoism, wrote "Tao Te Ching" ("The Book of the Way"). (600 BC–531 BC)

✧✧✧

Be careful what you water your dreams with. Water them with worry and fear and you will produce weeds that choke the life from your dream. Water them with optimism and solutions and you will cultivate success. Always be on the lookout for ways to turn a problem into an opportunity for success. Always be on the lookout for ways to nurture your dream.

- Lao Tzu (Tao-te Ching)

✧✧✧

Anticipate the difficult by managing the easy.

- Lao Tzu (Tao-te Ching)

✧✧✧

We put thirty spokes together and call it a wheel;
But it is on the space where there is nothing that the usefulness of the wheel depends.

We turn clay to make a vessel;
But it is on the space where there is nothing that the usefulness of the vessel depends.
We pierce doors and windows to make a house;
And it is on these spaces where there is nothing that the usefulness of the house depends.

Therefore just as we take advantage of what is, we should recognize the usefulness of what is not.

- Lao Tzu (Tao-te Ching)

✧✧✧

Knowing others is wisdom;
Knowing the self is enlightenment.
Mastering others requires force;
Mastering the self requires strength;
He who knows he has enough is rich.
Perseverance is a sign of will power.
He who stays where he is endures.
To die but not to perish is to be eternally present.

- Lao Tzu (Tao-te Ching)

✧✧✧

Learn as though you would never be able to master it; Hold it as though you would be in fear of losing it.

- Lao Tzu (Tao-te Ching)

✧✧✧

If you would take, you must first give, this is the beginning of intelligence.

- Lao Tzu (Tao-te Ching)

✧✧✧

To lead people, walk beside them... As for the best leaders, the people do not notice their existence. The next best, the people honor and praise. The next, the people fear; and the next, the people hate... When the best leader's work is done the people say, 'We did it ourselves!'

- Lao Tzu (Tao-te Ching)

✧✧✧

Do the difficult things while they are easy and do the great things while they are small. A journey of a thousand miles must begin with a single step.

- Lao Tzu (Tao-te Ching)

✧✧✧

One who is too insistent on his own views, finds few to agree with him.

- Lao Tzu (Tao-te Ching)

✧✧✧

It does not matter how slowly you go as long as you do not stop.

- Lao Tzu (Tao-te Ching)

✧✧✧

When it is obvious that the goals cannot be reached, don't adjust the goals, adjust the action steps.

- Lao Tzu (Tao-te Ching)

✧✧✧

Arms are instruments of ill omen. . . . When one is compelled to use them, it is best to do so without relish. There is no glory in victory, and to glorify it despite this is to exult in the killing of men. . . . When great numbers of people are killed, one should weep over them with sorrow. When victorious in war, one should observe mourning rites.

- Lao Tzu (Tao-te Ching)

✧✧✧

Better stop short than fill to the brim. Oversharpen the blade, and the edge will soon blunt. Amass a store of gold and jade, and no one can protect it. Claim wealth and titles, and disaster will follow. Retire when the work is done. This is the way of heaven.

- Lao Tzu (Tao-te Ching)

✧✧✧

Anticipate the difficult by managing the easy.

- Lao Tzu

✧✧✧

Do the difficult things while they are easy and do the great things while they are small. A journey of a thousand miles must begin with a single step.

- Lao Tzu

✧✧✧

Great acts are made up of small deeds.

- Lao Tzu

✧✧✧

He who conquers others is strong; He who conquers himself is mighty.

- Lao Tzu

✧✧✧

He who does not trust enough, will not be trusted.

- Lao Tzu

✧✧✧

The more laws and order are made prominent, the more thieves and robbers there will be.

- Lao Tzu

✧✧✧

War is a great affair of state, the realm of life and death, the road to safety or ruin, a thing to be studied with extreme diligence.

- Sun Tzu

✧✧✧

When the civil leadership is ignorant of military manoeuvres but shares equally in the command of armies, the soldiers get confused.

- Sun Tzu

✧✧✧

The art of war is of vital importance to the State. It is a matter of life and death, a road either to safety or to ruin. Hence it is a subject of inquiry which can on no account be neglected.

- Sun Tzu

✧✧✧

Learning the concentrated essence of military wisdom requires concreted minded soldiers to follow the course, counsel, teachings and evolve clarity of thought.

- Sun Tzu

✧✧✧

To rely on rustics and not prepare is the greatest of crimes;
To be prepared beforehand for any contingency is the greatest of virtues.

- Sun Tzu

✧✧✧

The quality of a decision is like the well-timed swoop of a falcon which enables it to strike and destroy its victim

- Sun Tzu

✧✧✧

All warfare is based on deception. Hence, when able to attack, we must seem unable; when using our forces, we must seem inactive; when we are near, we must make the enemy believe we are far away; when far away, we must make him believe we are near. Hold out baits to entice the enemy. Feign disorder, and crush him.

- Sun Tzu

✧✧✧

If your enemy is secure at all points, be prepared for him. If he is in superior strength, evade him. If your opponent is temperamental, seek to irritate him. Pretend to be weak, that he may grow arrogant. If he is taking his ease, give him no rest. If his forces are united, separate them. If sovereign and subject are in accord, put division between them. Attack him where he is unprepared, appear where you are not expected.

- Sun Tzu

✧✧✧

The clever combatant imposes his will on the enemy, but does not allow the enemy's will to be imposed on him.

- Sun Tzu

✧✧✧

There are three ways in which a ruler can bring misfortune upon his army;

(1) By commanding the army to advance or to retreat, being ignorant of the fact that it cannot obey. This is called hobbling the army.

(2) By attempting to govern an army in the same way as he administers a kingdom, being ignorant of the conditions which obtain in an army. This causes restlessness in the soldiers' mind.

(3) By employing the officers of his army without discrimination, through ignorance of the military principle of adaptation to circumstances. This shakes the confidence of the soldiers.

- Sun Tzu

✧✧✧

The art of using troops is this: ......When ten to the enemy's one, surround him; ......When five times his strength, attack him; ......If double his strength, divide him; ......If equally matched you may engage him; ......If weaker numerically, be capable of withdrawing; ......And if in all respects unequal, be capable of eluding him, ...for a small force is but booty for one more powerful.

- Sun Tzu (Art of War)

✧✧✧

The general who wins a battle makes many calculations in his temple ere the battle is fought. The general who loses a battle makes but few calculations beforehand. Thus do many calculations lead to victory, and few calculations to defeat: how much more no calculation at all! It is by attention to this point that I can foresee who is likely to w in or lose.

- Sun Tzu

✧✧✧

He who wishes to fight must first count the cost. When you engage in actual fighting, if victory is long in coming, then men's weapons will grow dull and their ardour will be dampened. If you lay siege to a town, you will exhaust your strength. Again, if the campaign is protracted, the resources of the State will not be equal to the strain. Now, when your weapons are dulled, your ardour dampened, your strength exhausted and your treasure spent, other chieftains will spring up to take advantage of your extremity. Then no man, however wise, will be able to avert the consequences that must ensue... In war, then, let your great object be victory, not lengthy campaigns.

- Sun Tzu

✧✧✧

Thus the highest form of generalship is to balk the enemy's plans, the next best is to prevent the junction of the enemy's forces, the next in order is to attack the enemy's army in the field, and the worst policy of all is to besiege walled cities.

- Sun Tzu

✧✧✧

Though we have heard of stupid haste in war, cleverness has never been seen associated with long delays.

- Sun Tzu

✧✧✧

It is only one who is thoroughly acquainted with the evils of war that can thoroughly understand the profitable way of carrying it on.

- Sun Tzu

✧✧✧

Bring war material with you from home, but forage on the enemy... use the conquered foe to augment one's own strength.

- Sun Tzu

✧✧✧

In the practical art of war, the best thing of all is to take the enemy's country whole and intact; to shatter and destroy it is not so good. So, too, it is better to recapture an army entire than to destroy it.

- Sun Tzu

✧✧✧

To fight and conquer in all your battles is not supreme excellence; supreme excellence consists in breaking the enemy's resistance without fighting.

- Sun Tzu

✧✧✧

If words of command are not clear and distinct, if orders are not thoroughly understood, the general is to blame. But if his orders are clear, and the soldiers nevertheless disobey, then it is the fault of their officers.

- Sun Tzu

✧✧✧

In the practical art of war, the best thing of all is to take the enemy's country whole and intact; to shatter and destroy it is not so good. So, too, it is better to recapture an army entire than to destroy it.

- Sun Tzu

✧✧✧

Victorious warriors win first and then go to war, while defeated warriors go to war first and then seek to win.

- Sun Tzu

✧✧✧

There are three ways in which a ruler can bring misfortune upon his army: By commanding the army to advance or to retreat, being ignorant of the fact that it cannot obey; This is called hobbling the army. By attempting to govern an army in the same way as he administers a kingdom, being ignorant of the conditions which obtain in an army; This causes restlessness in the soldier's minds. By employing the officers of his army without discrimination, through ignorance of the military principle of adaptation to circumstances. This shakes the confidence of the soldiers.

- Sun Tzu

✧✧✧

He will win who knows when to fight and when not to fight. He will win who knows how to handle both superior and inferior forces. He will win whose army is animated by the same spirit throughout all its ranks. He will win who, prepared himself, waits to take the enemy unprepared. He will win who has military capacity and is not interfered with by the sovereign.

- Sun Tzu

✧✧✧

If you know the enemy and know yourself, you need not fear the result of a hundred battles. If you know yourself but not the enemy, for every victory gained you will also suffer a defeat. If you know neither the enemy nor yourself, you will succumb in every battle.

- Sun Tzu

✧✧✧

The good fighters of old first put themselves beyond the possibility of defeat, and then waited for an opportunity of defeating the enemy. To secure ourselves against defeat lies in our own hands, but the opportunity of defeating the enemy is provided by the enemy himself. Thus the good fighter is able to secure himself against defeat, but cannot make certain of defeating the enemy.

- Sun Tzu

✧✧✧

Generally, he who occupies the battlefield first and awaits the enemy is at ease; He who comes later to the scene and rushes into the fight is weary.

- Sun Tzu

✧✧✧

Making no mistakes is what establishes the certainty of victory, for it means conquering an enemy that is already defeated.

- Sun Tzu

✧✧✧

The victorious strategist only seeks battle after the victory has been won, whereas he who is destined to defeat first fights and afterwards looks for victory.

- Sun Tzu

✧✧✧

Fighting with a large army under your command is nowise different from fighting with a small one: it is merely a question of instituting signs and signals.

- Sun Tzu

✧✧✧

An army may march great distances without distress, if it marches through country where the enemy is not. You can be sure of succeeding in your attacks if you only attack places which are undefended. You can ensure the safety of your defence if you only hold positions that cannot be attacked.

- Sun Tzu

✧✧✧

In all fighting, the direct method may be used for joining battle, but indirect methods will be needed in order to secure victory. In battle, there are not more than two methods of attack—the direct and the indirect; yet these two in combination give rise to an endless series of manoeuvres. The direct and the indirect lead on to each other in turn. It is like moving in a circle—you never come to an end. Who can exhaust the possibilities of their combination?

- Sun Tzu

✧✧✧

The clever combatant imposes his will on the enemy, but does not allow the enemy's will to be imposed on him.

- Sun Tzu

✧✧✧

Hence that general is skilful in attack whose opponent does not know what to defend; and he is skilful in defence whose opponent does not know what to attack.

- Sun Tzu

✧✧✧

If we wish to fight, the enemy can be forced to an engagement even though he be sheltered behind a high rampart and a deep ditch. All we need do is attack some other place that he will be obliged to relieve. If we do not wish to fight, we can prevent the enemy from engaging us even though the lines of our encampment be merely traced out on the ground. All we need do is to throw something odd and unaccountable in his way.

- Sun Tzu

✧✧✧

Should the enemy strengthen his van, he will weaken his rear; should he strengthen his rear, he will weaken his van; should he strengthen his left, he will weaken his right; should he strengthen his right, he will weaken his left. If he sends reinforcements everywhere, he will everywhere be weak.

- Sun Tzu

✧✧✧

In making tactical dispositions, the highest pitch you can attain is to conceal them.

- Sun Tzu

✧✧✧

Military tactics are like unto water; for water in its natural course runs away from high places and hastens downwards... Water shapes its course according to the nature of the ground over which it flows; the soldier works out his victory in relation to the foe whom he is facing. Therefore, just as water retains no constant shape, so in warfare there are no constant conditions. He who can modify his tactics in relation to his opponent and thereby succeed in winning, may be called a heaven-born captain.

- Sun Tzu

✧✧✧

So in war, the way is to avoid what is strong and to strike at what is weak.

- Sun Tzu

✧✧✧

The difficulty of tactical manoeuvring consists in turning the devious into the direct, and misfortune into gain.

- Sun Tzu

✧✧✧

Manoeuvring with an army is advantageous; with an undisciplined multitude, most dangerous.

- Sun Tzu

✧✧✧

We cannot enter into alliances until we are acquainted with the designs of our neighbours.

- Sun Tzu

✧✧✧

Do not interfere with an army that is returning home. When you surround an army, leave an outlet free. Do not press a desperate foe too hard.

- Sun Tzu

✧✧✧

The art of war teaches us to rely not on the likelihood of the enemy's not coming, but on our own readiness to receive him; not on the chance of his not attacking, but rather on the fact that we have made our position unassailable.

- Sun Tzu

✧✧✧

When the common soldiers are too strong and their officers too weak, the result is INSUBORDINATION. When the officers are too strong and the common soldiers too weak, the result is COLLAPSE. When the higher officers are angry and insubordinate, and on meeting the enemy give battle on their own account from a feeling of resentment, before the commander-in-chief can tell whether or no he is in a position to fight, the result is RUIN.

- Sun Tzu

✧✧✧

The general who advances without coveting fame and retreats without fearing disgrace, whose only thought is to protect his country and do good service for his sovereign, is the jewel of the kingdom.

- Sun Tzu

✧✧✧

On dispersive ground, therefore, fight not. On facile ground, halt not. On contentious ground, attack not. On open ground, do not try to block the enemy's way. On the ground of intersecting highways, join hands with your allies. On serious ground, gather in plunder. In difficult ground, keep steadily on the march. On hemmed-in ground, resort to stratagem. On desperate ground, fight.

- Sun Tzu

✧✧✧

Regard your soldiers as your children, and they will follow you into the deepest valleys; look upon them as your own beloved sons, and they will stand by you even unto death. If, however, you are indulgent, but unable to make your authority felt; kind-hearted, but unable to enforce your commands; and incapable, moreover, of quelling disorder: then your soldiers must be likened to spoilt children; they are useless for any practical purpose.

- Sun Tzu

✧✧✧

If we know that our own men are in a condition to attack, but are unaware that the enemy is not open to attack, we have gone only halfway towards victory. If we know that the enemy is open to attack, but are unaware that our own men are not in a condition to attack, we have gone only halfway towards victory. If we know that the enemy is open to attack, and also know that our men are in a condition to attack, but are unaware that the nature of the ground makes fighting impracticable, we have still gone only halfway towards victory.

- Sun Tzu

✧✧✧

If you know the enemy and know yourself, your victory will not stand in doubt; if you know Heaven and know Earth, you may make your victory complete.

- Sun Tzu

✧✧✧

If asked how to cope with a great host of the enemy in orderly array and on the point of marching to the attack, I should say: Begin by seizing something which your opponent holds dear; then he will be amenable to your will. Rapidity is the essence of war: take advantage of the enemy's unreadiness, make your way by unexpected routes, and attack unguarded spots.

- Sun Tzu

✧✧✧

Throw your soldiers into positions whence there is no escape, and they will prefer death to flight. If they will face death, there is nothing they may not achieve.

- Sun Tzu

✧✧✧

If our soldiers are not overburdened with money, it is not because they have a distaste for riches; if their lives are not unduly long, it is not because they are disinclined to longevity.

- Sun Tzu

✧✧✧

Bestow rewards without regard to rule, issue orders without regard to previous arrangements; and you will be able to handle a whole army as though you had to do with but a single man.

- Sun Tzu

✧✧✧

Unhappy is the fate of one who tries to win his battles and succeed in his attacks without cultivating the spirit of enterprise; for the result is waste of time and general stagnation. Hence the saying: The enlightened ruler lays his plans well ahead; the good general cultivates his resources.

- Sun Tzu

✧✧✧

Move not unless you see an advantage; use not your troops unless there is something to be gained; fight not unless the position is critical. If it is to your advantage, make a forward move; if not, stay where you are. Anger may in time change to gladness; vexation may be succeeded by content.

- Sun Tzu

✧✧✧

No leader should put troops into the field merely to gratify his ambition; no leader should fight a battle simply out of pique. But a kingdom that has once been destroyed can never come again into being; nor can the dead ever be brought back to life. Hence the enlightened leader is heedful, and the good leader full of caution.

- Sun Tzu

✧✧✧

Spies cannot be usefully employed without a certain intuitive sagacity; (2) They cannot be properly managed without benevolence and straight forwardness; (3) Without subtle ingenuity of mind, one cannot make certain of the truth of their reports; (4) Be subtle! be subtle! and use your spies for every kind of warfare; (5) If a secret piece of news is divulged by a spy before the time is ripe, he must be put to death together with the man to whom the secret was told.

- Sun Tzu

✧✧✧

The enemy's spies who have come to spy on us must be sought out, tempted with bribes, led away and comfortably housed. Thus they will become double agents and available for our service. It is through the information brought by the double agent that we are able to acquire and employ local and inward spies. It is owing to his information, again, that we can cause the doomed spy to carry false tidings to the enemy.

- Sun Tzu

✧✧✧

To capture the enemy's entire army is better than to destroy it; to take intact a regiment, a company, or a squad is better than to destroy them. For to win one hundred victories in one hundred battles is not the supreme of excellence. To subdue the enemy without fighting is the supreme excellence.

- Sun Tzu

✧✧✧

U

If you can find something everyone agrees on, it's wrong.

- Mo Udall

✧✧✧

We have, I fear, confused power with greatness.

- Stewart L. Udall
Commencement address, Dartmouth College, 13 June 1965

✧✧✧

The essence of a general's job is to assist in developing a clear sense of purpose to keep the junk from getting in the way of important things.

- Lt Gen Walter F. Ulmer, US Army

✧✧✧

Since wars begin in the minds of men, it is in the minds of men that the defences of peace must be constructed.

- UNESCO Charter

✧✧✧

Stoop and you'll be stepped on; stand tall and you'll be shot at.

- Carlos A. Urbizo

✧✧✧

Terrorism is the war of the poor. War is the terrorism of the rich.

- Leon Uris

✧✧✧

V

If God did not exist, it would be necessary to invent him.

-Voltaire (1694–1778)

François-Marie Arouet, better known by the pen name Voltaire

◆◆◆

It is dangerous to be right when the government is wrong.

-Voltaire

◆◆◆

Judge a man by his questions rather than by his answers.

-Voltaire

◆◆◆

The longest part of the journey is said to be the passing of the gate.

- Marcus Terentius Varro

◆◆◆

If you can't stand the heat, get out of the kitchen.

- Harry Vaughan

◆◆◆

It is not true that equality is a law of nature. Nature has no equality. Its sovereign law is subordination and dependence.

- Marquis de Vauvenargues

◆◆◆

To achieve great things we must live as though we were never going to die.

- Marquis de Vauvenargues

◆◆◆

Never tell everything at once.

- Ken Venturi, Two Great Rules of Life

◆◆◆

The prime purpose of eloquence is to keep other people from talking.

- Louis Vermeil

◆◆◆

I am absolutely, unalterably opposed to risking American lives for some sort of military and political objectives that we don't understand.

- General John Vessey, Chairman of the US Joint Chiefs, 1983

◆◆◆

It is not enough to succeed. Others must fail.

- Gore Vidal

✧✧✧

Today's public figures can no longer write their own speeches or books, and there is some evidence that they can't read them either.

- Gore Vidal

✧✧✧

Don't let it end like this. Tell them I said something.

- Pancho Villa, last words

✧✧✧

As a twig is bent the tree inclines.

- Virgil

✧✧✧

Look with favour upon a bold beginning.

- Virgil

✧✧✧

They can conquer who believe they can.

- Virgil

✧✧✧

The chief weapon of sea pirates, however, was their capacity to astonish. Nobody else could believe, until it was too late, how heartless and greedy they were.

- Kurt Vonnegut, Breakfast of Champions

✧✧✧

Another flaw in the human character is that everybody wants to build and nobody wants to do maintenance.

- Kurt Vonnegut, Hocus Pocus

✧✧✧

To acquire knowledge, one must study; but to acquire wisdom, one must observe.

- Marilyn vos Savant

✧✧✧

Ecce Homo

Behold the man.

- Vulgate

✧✧✧

W

By the time a man realizes that maybe his father was right, he usually has a son who thinks he's wrong.

- Charles Wadsworth

✧✧✧

Joy is not in things; it is in us.

- Richard Wagner

✧✧✧

The terrible thing about terrorism is that ultimately it destroys those who practise it. Slowly but surely, as they try to extinguish life in others, the light within them dies.

-Terry Waite, London Guardian, Feb. 20, 1992

✧✧✧

Most people assume the fights are going to be the left versus the right, but it always is the reasonable versus the jerks.

- Jimmy Wales, Founder of Wikipedia

✧✧✧

Why does the Air Force need expensive new bombers? Have the people we've been bombing over the years been complaining?

- George Wallace

✧✧✧

To be one's self, and unafraid whether right or wrong, is more admirable than the easy cowardice of surrender to conformity.

- Irving Wallace

✧✧✧

With freedom comes responsibility, and I can think of no responsibility greater than putting on a U.S military uniform and standing in the gap between an enemy threat and civilian life.

- Zack Wamp

✧✧✧

The excellence of a gift lies in its appropriateness rather than in its value.

- Charles Dudley Warner, 1873

✧✧✧

One man cannot hold another man down in the ditch without remaining down in the ditch with him.

- Booker T. Washington

❖❖❖

Success is to be measured not so much by the position that one has reached in life as by the obstacles which he has overcome.

- Booker T. Washington

❖❖❖

There are two ways of exerting one's strength: one is pushing down, the other is pulling up.

- Booker T. Washington

❖❖❖

It is better to offer no excuse than a bad one.

- George Washington (1732–1799)
First President of the United States (1789–1797)

❖❖❖

Few men have virtue to withstand the highest bidder.

- George Washington

❖❖❖

Associate with men of good quality if you esteem your own reputation; for it is better to be alone than in bad company.

- George Washington

❖❖❖

Be courteous to all, but intimate with few, and let those few be well tried before you give them your confidence.

- George Washington

❖❖❖

Discipline is the soul of an army. It makes small numbers formidable; procures success to the weak, and esteem to all.

- George Washington

❖❖❖

If we desire to avoid insult, we must be able to repel it; if we desire to secure peace, one of the most powerful instruments of our rising prosperity, it must be known, that we are at all times ready for War.

- George Washington

❖❖❖

To be prepared for war is one of the most effective means of preserving peace.

- George Washington

❖❖❖

Experience teaches us that it is much easier to prevent an enemy from posting themselves than it is to dislodge them after they have got possession.

- George Washington

❖❖❖

My observation is that whenever one person is found adequate to the discharge of a duty... it is worse executed by two persons, and scarcely done at all if three or more are employed therein.

- George Washington

✧✧✧

Nothing can be more hurtful to the service, than the neglect of discipline; for that discipline, more than numbers, gives one army the superiority over another.

- George Washington

✧✧✧

The reason most people never reach their goals is that they don't define them, or ever seriously consider them as believable or achievable. Winners can tell you where they are going, what they plan to do along the way, and who will be sharing the adventure with them.

- Denis Watley

✧✧✧

Again, speed in action must be cultivated; the power to think quickly in an emergency is one of the greatest assets both of the boxer and the commander; and the power to move quickly often gives to a body of troops, as to a boxer, the advantage of surprise.

- Field Marshal A. Wavell
(The Good Soldier)

✧✧✧

Efficiency in a general his soldiers have a right to expect; geniality they are usually right to suspect.

- Field Marshal A. Wavell
(The Good Soldier)

✧✧✧

Good generals, unlike poets, are made rather than born, and will never reach the first rank without much study of their profession; but they must have certain natural gifts-the power of quick decision, judgment, boldness, and, I am afraid, a considerable degree of toughness, almost callousness, which is harder to find as civilisation progresses.

- Field Marshal A. Wavell
(The Good Soldier)

✧✧✧

The chief assets of the attacker are: firstly, the feeling of moral superiority which an offensive attitude gives, similar to that of the boxer who carries the fight to his opponent; and, secondly, the power to choose at what place, by what method, and at what time the main action will be fought.

- Field Marshal A. Wavell
(The Good Soldier)

✧✧✧

Let us be clear about three facts: First, all battles and all wars are won in the end by the infantryman. Secondly, the infantryman always bears the brunt. His casualties are heavier, he suffers greater extremes of discomfort and fatigue than the other arms. Thirdly, the art of the infantryman is less stereotyped and far harder to acquire in modern war than that of any other arm.

- Field-Marshal Earl Wavell

✧✧✧

No amount of study or learning will make a man a leader unless he has the natural qualities of one.

- Field-Marshal Earl Wavell

✧✧✧

The best soldier has in him, I think, a seasoning of devilry.

- Field-Marshal Earl Wavell

✧✧✧

Winston Churchill is always expecting rabbits to come out of an empty hat.

- Field-Marshal Earl Wavell

✧✧✧

Military history is a flesh-and-blood affair, and not a matter of diagrams and formulas or of rules; not a conflict of machines but of men. In the lecture hall of a French Infantry School which I once attended was written the following from Ardant du Picq: "The man is the first weapon of battle: let us then study the soldier in battle, for it is he who brings reality to it. Only the study of the past can give us a sense of reality, and show us how the soldier will fight in battle." When you study Military History, don't read Outlines on strategy or the principles of war. Read biographies, memoirs historical novels...Get at the flesh and blood of it; not the skeleton.

- Field-Marshal Earl Wavell, Generals and Generalship

✧✧✧

Power is the possibility of imposing one's will upon the behaviour of other persons.

- Max Weber, German sociologist (1864-1920)

✧✧✧

The fate of our times is characterised by rationalisation and intellectualisation and, above all, by the "disenchantment of the world.

- Max Weber

✧✧✧

Bureaucratic administration means fundamentally domination through knowledge.

- Max Weber

✧✧✧

The decisive reason for the advance of the bureaucratic organisation has always been its purely technical superiority over any other form of organisation.

- Max Weber

✧✧✧

Be discreet in all things, and so render it unnecessary to be mysterious about any.

- Arthur Wellesley
(first Duke of Wellington)

✧✧✧

Publish and be damned.

- Field Marshal Arthur Wellesley
Duke of Wellington (1769–1852)

✧✧✧

In 1824 Wellington received a letter from a publisher offering to refrain from issuing an edition of the rather racy memoirs of one of his mistresses, Harriette Wilson, in exchange for financial consideration. The Duke promptly returned the missive, after scrawling these words across it.

I don't know what effect these men will have upon the enemy, but, by God, they frighten me.

- Arthur Wellesley
(first Duke of Wellington)

✧✧✧

Nothing except a battle lost can be half as melancholy as a battle won.

- Arthur Wellesley
(first Duke of Wellington)

✧✧✧

All the business of war, and indeed all the business of life, is to endeavor to find out what you don't know by what you do; that's what I called 'guessing what was at the other side of the hill.'

- Arthur Wellesley
(first Duke of Wellington)

✧✧✧

It has been a damned serious business—Blücher and I have lost 30,000 men. It has been a damned nice thing—the nearest run thing you ever saw in your life...By God! I don't think it would have done if I had not been there.

- Arthur Wellesley
(first Duke of Wellington)

✧✧✧

Wise people learn when they can; fools learn when they must.

- Arthur Wellesley
(first Duke of Wellington)

✧✧✧

As Lord Chesterfield said of the generals of his day, 'I only hope that when the enemy reads the list of their names, he trembles as I do.'

- Arthur Wellesley
(first Duke of Wellington)

✧✧✧

My Lord,

If I attempted to answer the mass of futile correspondence which surrounds me, I should be debarred from the serious business of campaigning... So long as I retain an independent position, I shall see no officer under my command is debarred by attending to the futile drivelling of mere quill-driving from attending to his first duty, which is and always has been to train the private men under his command that they may without question beat any force opposed to them in the field.

- Arthur Wellesley, (first Duke of Wellington)
To the Secretary of State for War during the Peninsular Campaign

✧✧✧

Your true value depends entirely on what you are compared with.

- Bob Wells

✧✧✧

They realized that the chief danger in aerial warfare from an excitable and intelligent public would be a clamour for local airships and aeroplanes to defend local interests.

- H.G. Wells (The War in the Air)

✧✧✧

And in the air are no streets, no channels, no point where one can say of an antagonist, "If he wants to reach my capital he must come by here." In the air all directions lead everywhere.

- H.G. Wells (The War in the Air)

✧✧✧

Before a war military science seems a real science, like astronomy; but after a war it seems more like astrology.

- Rebecca West

✧✧✧

War is fear cloaked in courage.

- General William Westmoreland
(1914–2005) United States Army

✧✧✧

As the senior commander in Vietnam, I was aware of the potency of public opinion—and I worried about it.

- General William Westmoreland

✧✧✧

I do not believe that the men who served in uniform in Vietnam have been given the credit they deserve. It was a difficult war against an unorthodox enemy.

- General William Westmoreland

✧✧✧

I don't think I have been loved by my troops, but I think I have been respected.

- General William Westmoreland

✧✧✧

It's the first war we've ever fought on the television screen and the first war that our country ever fought where the media had full reign.

- General William Westmoreland

✧✧✧

Militarily, we succeeded in Vietnam. We won every engagement we were involved in out there.

- General William Westmoreland

✧✧✧

The military don't start wars. Politicians start wars.

- General William Westmoreland

✧✧✧

Successful organizations, including the Military, have learned that the higher the risk, the more necessary it is to engage everyone's commitment and intelligence.

- Margaret J. Wheatley

✧✧✧

We think in generalities, but we live in detail.

- Alfred North Whitehead

✧✧✧

Once we commit force, we must be prepared to back it up as opposed to just sending soldiers into operations for limited goals.

- General John Wickham,
US Army Chief of Staff

✧✧✧

Fighting terrorism is like being a goalkeeper. You can make a hundred brilliant saves but the only shot that people remember is the one that gets past you.

- Paul Wilkinson, London Daily Telegraph,
1 September 1992

✧✧✧

Of all the disasters of Vietnam, the worst may be the "lessons" that we'll draw from it. . . . Lessons from such complex events require much reflection to be of more than negative worth. But reactions to Vietnam . . . tend to be visceral rather than reflective.

- Albert Wohlstetter

✧✧✧

I not only use all the brains that I have, but all that I can borrow.

- Woodrow Wilson

✧✧✧

The man who is swimming against the stream knows the strength of it.

- Woodrow Wilson

✧✧✧

You cannot be friends upon any other terms than upon the terms of equality.

- Woodrow Wilson

✧✧✧

THE BURIAL OF SIR JOHN MOORE AT CORUNNA

Not a drum was heard, nor a funeral note,
As his corpse to the rampart we hurried;
Not a soldier discharged his farewell shot
O'er the grave where our hero we buried.

We buried him darkly at dead of night,
The sods with our bayonets turning;
By the struggling moonbeam's misty light
And the lantern dimly burning.

No useless coffin enclosed his breast,
Nor in sheet nor in shroud we wound him;
But he lay like a warrior taking his rest
With his martial cloak around him.

Few and short were the prayers we said,
And we spoke not a word of sorrow;
But we steadfastly gazed on the face that was dead,
And we bitterly thought of the morrow.

We thought, as we hollowed his narrow bed
And smoothed down his lonely pillow,
That the foe and the stranger would tread o'er his head,
And we far away on the billow!

Lightly they'll talk of the spirit that's gone
And o'er his cold ashes upbraid him,—
But little he'll wreck, if they let him sleep on
In the grave where a Briton has laid him.

But half of our heavy task was done
When the clock struck the hour for retiring:
And we heard the distant and random gun
That the foe was sullenly firing.

Slowly and sadly we laid him down,
From the field of his fame fresh and gory;
We carved not a line, and we raised not a stone,
But left him alone with his glory.

- Charles Wolfe (1791–1823)

✧✧✧

Possessed with a full confidence of the certain success which British valor must gain over such enemies, I have led you up these steep and dangerous rocks, only solicitous to show you the foe within your reach.

- Major General James P. Wolfe (1727–1759)
British Army

✧✧✧

The impossibility of a retreat makes no difference in the situation of men resolved to conquer or die; and, believe me, my friends, if your conquest could be bought with the blood of your general, he would most cheerfully resign a life which he has long devoted to his country.

- Major General James P. Wolfe

✧✧✧

The impossibility of a retreat makes no difference in the situation of men resolved to conquer or die; and, believe me, my friends, if your conquest could be bought with the blood of your general, he would most cheerfully resign a life which he has long devoted to his country.

- Major General James P. Wolfe

✧✧✧

You know too well the forces which compose their army to dread their superior numbers.

- Major General James P. Wolfe

✧✧✧

Illusion is an anodyne, bred by the gap between wish and reality.

- Herman Wouk

✧✧✧

Peace, if it ever exists, will not be based on the fear of war, but on the love of peace. It will not be the abstaining from an act, but the coming of a state of mind.

- Herman Wouk
(The Winds of War)

✧✧✧

For the rest of us, perhaps. Not for the dead, not for the more than fifty million real dead in the world's worst catastrophe: victors and vanquished, combatants and civilians, people of so many nations, men, women, and children, all cut down. For them there can be no new earthly dawn. Yet thought their bones like in the darkness of the grave, they will not have died in vain, if their remembrance can lead us from the long, long time of war to the time for peace.

- Herman Wouk
(War and Remembrance)

✧✧✧

X

Power never takes a back step—only in the face of more power.

- Malcolm X
Malcolm X Speaks, 1965

✧✧✧

The art of war is, in the last result, the art of keeping one's freedom of action

- Xenophon, Greek historian
(c. 430-355 BC)

✧✧✧

It doesn't matter if a cat is black or white, so long as it catches mice.

- Deng Xiaoping

✧✧✧

Observe calmly; secure our position; cope with affairs calmly; hide our capacities; and bide our time.

- Deng Xiaoping
24 Character Strategy

✧✧✧

Y

I was always afraid of dying. Always. It was my fear that made me learn everything I could about my airplane and my emergency equipment, and kept me flying respectful of my machine and always alert in the cockpit.

- Major General Chuck Yeager, US Air Force

✧✧✧

If you want to grow old as a pilot, you've got to know when to push it, and when to back off.

- Major General Chuck Yeager, US Air Force

✧✧✧

Most pilots learn, when they pin on their wings and go out and get in a fighter, especially, that one thing you don't do, you don't believe anything anybody tells you about an airplane.

- Major General Chuck Yeager, US Air Force

✧✧✧

Never wait for trouble.

- Major General Chuck Yeager, US Air Force

✧✧✧

You don't concentrate on risks. You concentrate on results. No risk is too great to prevent the necessary job from getting done.

- Major General Chuck Yeager, US Air Force

✧✧✧

I have flown in just about everything, with all kinds of pilots in all parts of the world—British, French, Pakistani, Iranian, Japanese, Chinese—and there wasn't a dime's worth of difference between any of them except for one unchanging, certain fact: the best, most skilful pilot has the most experience.

- Major General Chuck Yeager, US Air Force

✧✧✧

Think like a wise man but communicate in the language of the people.

- William Butler Yeats

✧✧✧

Procrastination is the thief of time.

- Edward Young (1683–1765)

✧✧✧

Peace of mind is that mental condition in which you have accepted the worst.

- Lin Yutang

✧✧✧

Besides the noble art of getting things done, there is a nobler art of leaving things undone. The wisdom of life consists in the elimination of nonessentials.

- Lin Yutang

✧✧✧

Z

The reason we have two ears and only one mouth, is that we may hear more and speak less.

- Zeno (335 BC–264 BC)
Greek philosopher

✧✧✧

It is necessary to investigate both the facts and the history of a problem in order to study and understand it.

- Chairman Mao Zedong

✧✧✧

An army that is cherished and respected by the people, and vice versa, is a nearly invincible force. The army and the people must unite on the grounds of basic respect.

- Chairman Mao Zedong

✧✧✧

Every Communist must grasp the truth, "Political power grows out of the barrel of a gun."

- Chairman Mao Zedong

✧✧✧

We are advocates of the abolition of war, we do not want war; but war can only be abolished through war, and in order to get rid of the gun, it is necessary to take up the gun.

- Chairman Mao Zedong

✧✧✧

Just because we have won a victory, we must never relax our vigilance against the frenzied plots for revenge by the imperialists and their running dogs. Whoever relaxes vigilance will disarm himself politically and land himself in a passive position.

- Chairman Mao Zedong

✧✧✧

The army must become one with the people so that they see it as their own army. Such an army will be invincible.

- Chairman Mao Zedong

✧✧✧

The commanders and fighters of the entire Chinese People's Liberation Army absolutely must not relax in the least their will to fight; any thinking that relaxes the will to fight and belittles the enemy is wrong.

- Chairman Mao Zedong

✧✧✧

But when it moves to the peaceful years I hate it. Not because I love chaos but because a time of peace is not good for the development of the people. It is unbearable.

- Chairman Mao Zedong

✧✧✧

We should rid our ranks of all impotent thinking. All views that overestimate the strength of the enemy and underestimate the strength of the people are wrong.

- Chairman Mao Zedong

✧✧✧

Without preparedness superiority is not real superiority and there can be no initiative either. Having grasped this point, a force, which is inferior but prepared can often defeat a superior enemy by surprise attack.

- Chairman Mao Zedong

✧✧✧

Who are our enemies? Who are our friends? This is a question of the first importance for the revolution.

- Chairman Mao Zedong

✧✧✧

Many people think it impossible for guerrillas to exist for long in the enemy's rear. Such a belief reveals lack of comprehension of the relationship that should exist between the people and the troops. The former may be likened to water the latter to the fish who inhabit it. How may it be said that these two cannot exist together?

- Chairman Mao Zedong

✧✧✧

Politics is war without bloodshed, while war is politics with bloodshed.

- Chairman Mao Zedong

✧✧✧

War can only be abolished by war, and in order to get rid of the gun it is necessary to take up the gun.

- Chairman Mao Zedong

✧✧✧

If you want to know the taste of a pear, you must change the pear by eating it yourself. . . . If you want to know the theory and methods of revolution, you must take part in revolution. All genuine knowledge originates in direct experience.

- Chairman Mao Zedong

✧✧✧

Weapons are an important factor in war, but not the decisive one; it is man and not materials that counts.

- Chairman Mao Zedong

✧✧✧

An army without culture is a dull-witted army, and a dull-witted army cannot defeat the enemy.

- Chairman Mao Zedong

✧✧✧

When the enemy advances, withdraw; when he stops, harass; when he tires, strike; when he retreats, pursue.

- Chairman Mao Zedong

✧✧✧

Be resolute, fear no sacrifice, and surmount every difficulty to win victory.

- Chairman Mao Zedong

✧✧✧

What is a constitution? It is a booklet with twelve or ten pages. I can tear them away and say that tomorrow we shall live under a different system. Today, the people will follow wherever I lead. All the politicians including the once mighty Mr. Bhutto will follow me with tails wagging.

- General Muhammad Zia-ul-Haq (1924–1988)

✧✧✧

I genuinely feel that the survival of this country lies in democracy and democracy alone.

- General Muhammad Zia-ul-Haq (1924–1988)

✧✧✧

Terrorism has replaced Communism as the rationale for the militarization of the country [America], for military adventures abroad, and for the suppression of civil liberties at home. It serves the same purpose, serving to create hysteria.

- Howard Zinn, Terrorism and War

✧✧✧

It's not right to respond to terrorism by terrorizing other people. And furthermore, it's not going to help. Then you might say, "Yes, it's terrorizing people, but it's worth doing because it will end terrorism." But how much common sense does it take to know that you cannot end terrorism by indiscriminately dropping bombs?

- Howard Zinn, Terrorism and War

✧✧✧

It might be interesting to wonder why all the generals see it in the same way, and all those, who never fired a shot in anger and really held back to go to war, see it in a different way. That's usually the way it is in history.

- General Anthony Zinni

✧✧✧

Diplomacy is a continuation of war by other means.

- Zhou En Lai

✧✧✧

Spirit must be transformed into material force before it can move the world forward.

- Zhou En Lai

✧✧✧